Solutions Manual to Accompany

Advanced Strength and Applied Elasticity

Third Edition

A. C. Ugural

New Jersey Institute of Technology

S. K. Fenster

New Jersey Institute of Technology

Prepared by A. C. Ugural

P T R Prentice-Hall, Englewood Cliffs, New Jersey 07632

Prentice-Hall, Inc.
A Paramount Communications Company
Englewood Cliffs, NJ 07632

Printed in the United States of America
10 9 8 7 6 5 4 3 2

ISBN 0-13-294703-X

Prentice-Hall International (UK) Limited, *London*
Prentice-Hall of Australia Pty. Limited, *Sydney*
Prentice-Hall Canada Inc., *Toronto*
Prentice-Hall Hispanoamericana, S.A., *Mexico*
Prentice-Hall of India Private Limited, *New Delhi*
Prentice-Hall of Japan, Inc., *Tokyo*
Simon & Schuster Asia Pte. Ltda., *Singapore*
Editora Prentice-Hall do Brasil, Ltda., *Rio de Janeiro*

CONTENTS

1.1

We have

$A=50\times75=3.75(10^{-3})\ m^2$
$\theta=50^\circ$ and $\sigma_x=P/A$.

Equations (1.8), with $\theta=50^\circ$:

$$\sigma_{x'}=700(10^3)=\sigma_x\cos^2 50^\circ=0.413\sigma_x =110.18P$$

or

$$P=6.35\ kN$$

$$|\tau_{x'y'}|=560(10^3)=\sigma_x\sin50^\circ\cos50^\circ =0.492\sigma_x=131.2P$$

or

$$P=4.27\ kN$$

Thus

$$P_{all}=4.27\ kN$$ ◄

1.2

Normal stress is

$$\sigma_x=\frac{P}{A}=\frac{125(10^3)}{0.05\times0.05}=50\ MPa$$

(a) Equations (1.8), with $\theta=20^\circ$:

$$\sigma_{x'}=50\cos^2 20^\circ=44.15\ MPa$$

$$\tau_{x'y'}=-50\sin20^\circ\cos20^\circ=-16.08\ MPa$$ ◄

$$\sigma_{y'}=50\cos^2(20^\circ+90^\circ)=5.849\ MPa$$

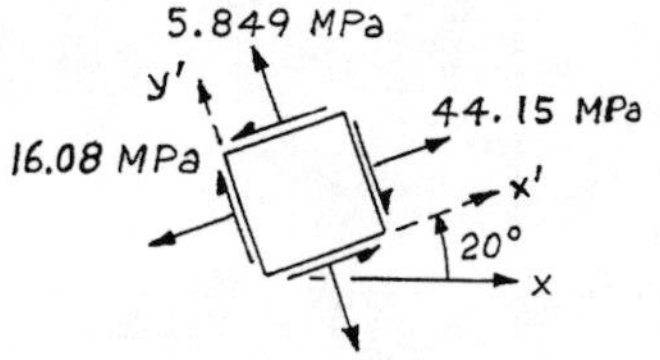

(b) Equations (1.8), with $\theta=45^\circ$:

$$\sigma_{x'}=50\cos^2 45^\circ=25\ MPa$$

$$\tau_{x'y'}=-50\sin45^\circ\cos45^\circ=-25\ MPa$$ ◄

$$\sigma_{y'}=50\cos^2(45^\circ+90^\circ)=25\ MPa$$

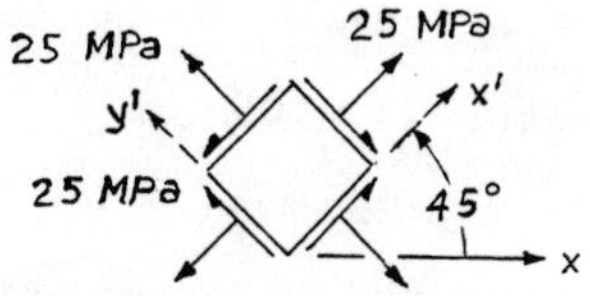

1.3

Refer to Fig. 1.6b.

Equations (1.8) by substituting the double angle-trigonometric relations, or Eqs. (1.14) with $\sigma_y=0$ and $\tau_{xy}=0$, become

$$\sigma_{x'}=\frac{1}{2}\sigma_x+\frac{1}{2}\sigma_x\cos2\theta$$

$$|\tau_{x'y'}|=\frac{1}{2}\sigma_x\sin2\theta$$

or

$$20=\frac{P}{2A}(1+\cos2\theta)$$

$$10=\frac{P}{2A}\sin2\theta$$

The foregoing leads to

$$2\sin2\theta-\cos2\theta=1 \qquad (a)$$

By introducing trigonometric identities, Eq. (a) becomes

$$4\sin\theta\cos\theta-2\cos^2\theta=0$$

or, $\tan\theta=1/2$,

$$\theta=26.56^\circ$$ ◄

Thus,

$$20=\frac{P}{2(1300)}(1+0.6)$$

gives

$$P=32.5\ kN$$ ◄

It can be shown that use of Mohr's circle yields readily the same result.

1.4

$$\theta=40+90=130^\circ$$

$$\sigma_x=\frac{P}{A}=-\frac{150(10^3)}{\pi(0.08^2-0.07^2)}=-31.83\ MPa$$

Equations (1.8):

$$\sigma_{x'}=-31.83\cos^2 130^\circ=-13.15\ MPa$$ ◄

$$\tau_{x'y'}=31.83\sin130^\circ\cos130^\circ =-15.67\ MPa$$ ◄

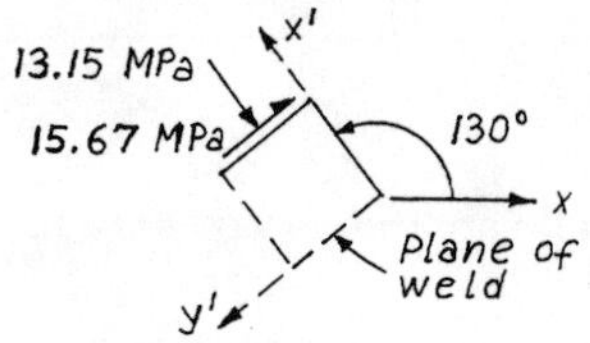

1.5

Use Eqs. (1.11),

$$(2x)+(-2xy)+(x)+F_x=0$$

$$(-y^2)+(-2yz+x)+(0)+F_y=0$$

$$(z-4xy)+(0)+(-2z)+F_z=0$$

Solving, we have (in MN/m^3):

$$F_x=-3x+2xy$$

$$F_y=-x+y^2+2yz \qquad (a)$$

$$F_z=4xy+z$$

Substituting x=-0.01 m, y=0.03 m, and z=0.06 m, Eqs. (a) yield the following values

$F_x=29.4$ kN/m³, $F_y=14.5$ kN/m³

$F_z=58.8$ kN/m³

Resultant body force is thus

$$F=\sqrt{F_x^2+F_y^2+F_z^2}$$

$$=67.32 \text{ kN/m}^3$$

1.6

Equations (1.11):

$$-2c_1y-2c_1y+0+0=0, \qquad 4c_1y\neq 0$$
$$0+c_3z+0+0=0, \qquad c_3z\neq 0$$
$$0+0+0+0=0$$

No. Eqs. (1.11) are not satisfied.

1.7

(a) No. Eqs. (1.11) are not satisfied.
(b) Yes. Eqs. (1.11) are satisfied.

1.8

Eqs. (1.11) for the given stress field yield

$$F_x=F_y=F_z=0$$

1.9

(a) From Mohr's circle, Fig. (a):

$\sigma_1=121$ MPa $\quad \sigma_2=-71$ MPa
$\tau_{max}=96$ MPa

$\theta_p'=-19.3°$

$\theta_s'=25.7°$

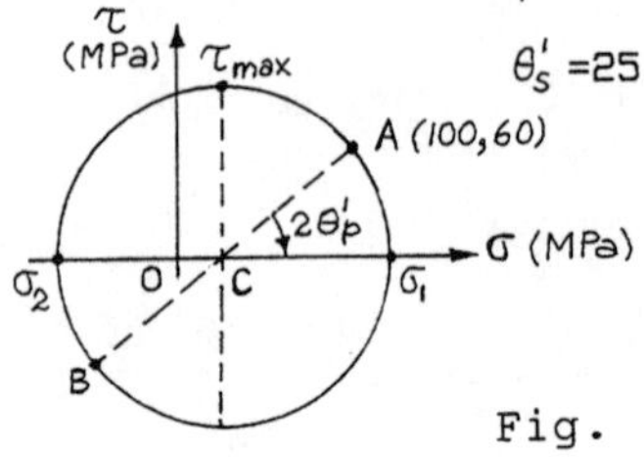

Fig. (a)

By applying Eq. (1.16):

$$\sigma_{1,2}=\frac{50}{2}\pm\left[\frac{22{,}500}{4}+3600\right]^{1/2}=25\pm 96$$

or

$\sigma_1=121$ MPa $\quad \sigma_2=-71$ MPa

Using Eq. (1.15):

$$\tan 2\theta_p=-\frac{12}{15}=-0.8$$

$\theta_p'=-19.3°$ $\quad \theta_s'=25.7°$

(b) From Mohr's circle, Fig. (b):

$\sigma_1=200$ MPa $\quad \sigma_2=-50$ MPa
$\tau_{max}=125$ MPa

$\theta_p'=26.15°$

$\theta_s'=71.55°$

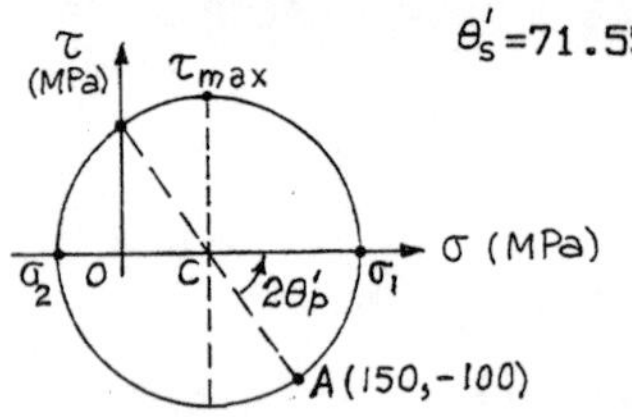

Fig. (b)

Using Eq. (1.16),

$$\sigma_{1,2}=75\pm\left[\frac{22{,}500}{4}+10{,}000\right]^{1/2}=75\pm 125$$

or

$\sigma_1=200$ MPa $\quad \sigma_2=-50$ MPa

Equation (1.15) gives

$$\tan 2\theta_p=4/3$$

$\theta_p'=26.55°$ $\quad \theta_s'=71.55°$

1.10

Referring to Mohr's circle, Fig. 1.13:

$$\sigma_{x'} = \frac{\sigma_1+\sigma_2}{2} + \frac{\sigma_1-\sigma_2}{2}\cos 2\theta$$
$$\sigma_{y'} = \frac{\sigma_1+\sigma_2}{2} - \frac{\sigma_1-\sigma_2}{2}\cos 2\theta \qquad \text{(a)}$$
$$\tau_{x'y'} = \frac{\sigma_1-\sigma_2}{2}\sin 2\theta \qquad \text{(b)}$$

From Eqs. (a),

$$\sigma_{x'} + \sigma_{y'} = \sigma_1 + \sigma_2$$

By using $\cos^2 2\theta+\sin^2 2\theta=1$, and Eqs. (a) and (b), we have

$$\sigma_{x'}\cdot\sigma_{y'} - \tau_{x'y'}^2 = \sigma_1\cdot\sigma_2 = \text{const.} \quad ◄$$

1.11

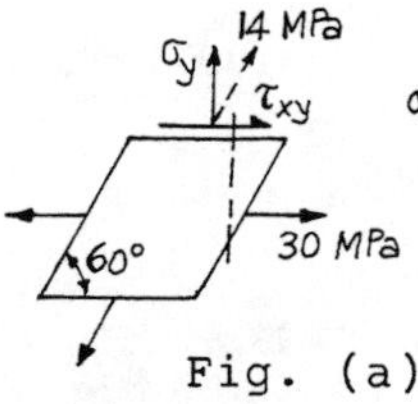

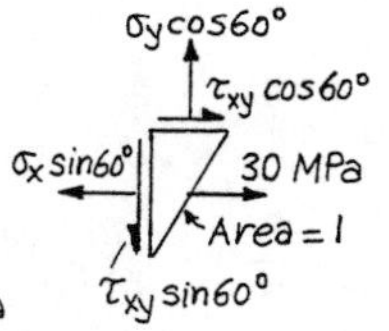

Fig. (a) Fig. (b)

(a) Figure (a):

$$\sigma_y = 14 \sin 60° = 12.12 \text{ MPa}$$
$$\tau_{xy} = 14 \cos 60° = 7 \text{ MPa} \quad ◄$$

Figure (b):

$$\Sigma F_y = 12.12 \cos 60° - \tau_{xy}\sin 60° = 0$$

or

$$\tau_{xy} = 7 \text{ MPa} \quad \text{(as before)}$$
$$\Sigma F_x = -\sigma_x \sin 60° + 30 + 7\cos 60° = 0$$

or

$$\sigma_x = 38.68 \text{ MPa} \quad ◄$$

(b) Equation (1.16) is therefore:

$$\sigma_{1,2} = \frac{38.68+12.12}{2} \pm \left[\left(\frac{38.68-12.12}{2}\right)^2 + 7^2\right]^{1/2}$$

or

$$\sigma_1 = 40.41 \text{ MPa}, \quad \sigma_2 = 10.39 \text{ MPa} \quad ◄$$

Also,

$$\theta_p = \frac{1}{2}\tan^{-1}\frac{2(7)}{38.68-12.12} = 13.9°$$

Note: Eq. (1. 14a) gives, $\sigma_{x'}$= 40.41 MPa. Thus,

$$\theta_p' = 13.9° \quad ◄$$

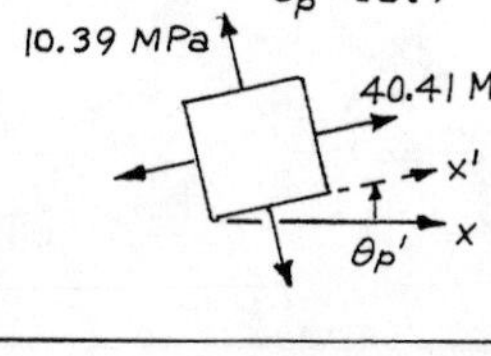

1.12

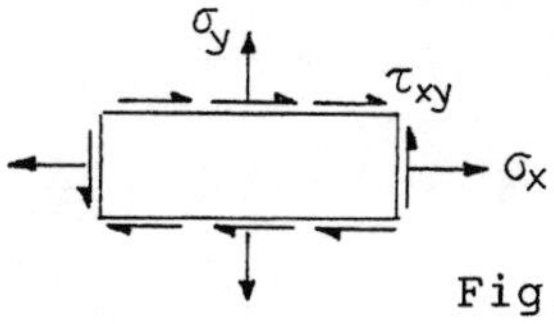

Fig. (a)

Figure (a):

$$\sigma_x = 100 \cos 45° = 70.7 \text{ MPa}$$
$$\sigma_y = 100 \sin 45° = 70.7 \text{ MPa}$$
$$\tau_{xy} = 100 \cos 45° = 70.7 \text{ MPa}$$

Now, Eqs. (1.14) give (Fig. b):

$$\sigma_{x'} = 70.7+0+70.7 \sin 240° = 9.47 \text{ MPa}$$
$$\tau_{x'y'} = -0+70.7 \cos 240° = -35.35 \text{ MPa} \quad ◁$$
$$\sigma_{y'} = 70.7-0-70.7 \sin 240° = 131.9 \text{ MPa}$$

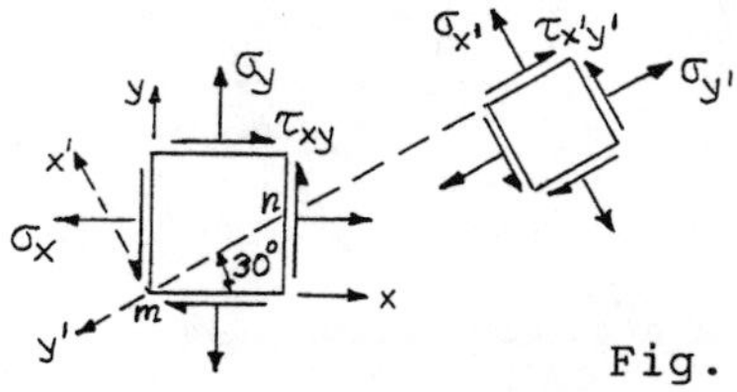

Fig. (b)

1.13

$$\sigma_y = -70 \sin 30° = -35 \text{ MPa}$$
$$\tau_{xy} = 70 \cos 30° = 60.6 \text{ MPa} \quad ◄$$

(a) Figure (a):

$$\Sigma F_x = -150 + 0.5\sigma_x + 60.6(0.866) = 0$$

or $\sigma_x = 195$ MPa

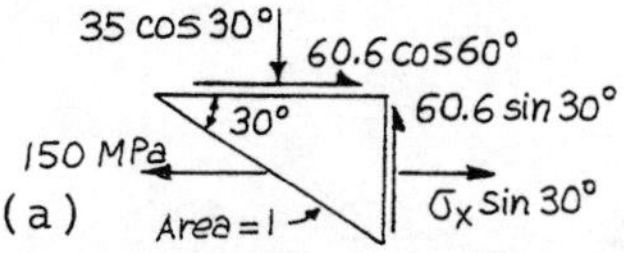

Fig. (a)

(b) Equation (1.16):

$$\sigma_{1,2} = \frac{195-35}{2} \pm \left[\left(\frac{195+35}{2}\right)^2 + 60.6^2\right]^{1/2}$$

or

$$\sigma_1 = 210 \text{ MPa}, \quad \sigma_2 = -50 \text{ MPa} \quad ◄$$

Also,

$$\theta_p = \frac{1}{2}\tan^{-1}\frac{2(60.6)}{195+35} = 13.89°$$

Equation (1.14a):

$$\sigma_{x'} = 80 + 115 \cos 2(13.89°) + 60.6 \sin 2(13.89°) = 210 \text{ MPa}$$

Thus,

$$\theta_p' = 13.89° \quad ◄$$

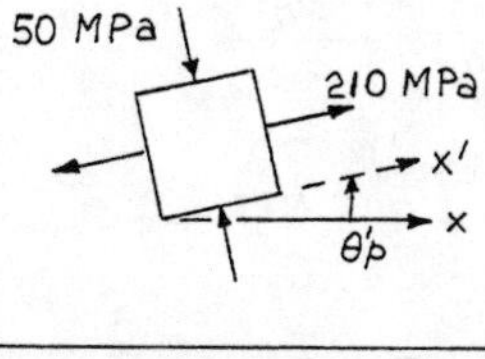

1.14

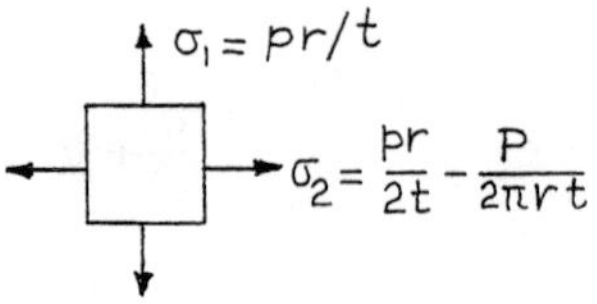

For pure shear, $\sigma_1 = -\sigma_2$:

$$\frac{pr}{t} = -\frac{pr}{2t} + \frac{P}{2\pi rt}$$

from which

$$P = 3\pi pr^2 \quad ◁$$

1.15

Table C.1:

$A = 2\pi rt$

$J = 2\pi r^3 t$

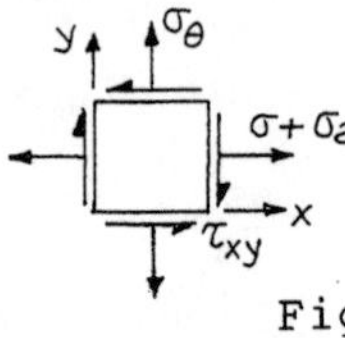

Fig. (a)

Stresses are (Fig. a):

$$\sigma = \frac{-P}{A} = \frac{-30(10^3)}{2\pi(0.12)(0.005)} = -25 \text{ MPa}$$

$$\sigma_a = \frac{pr}{2t} = \frac{4(10^6)\,120}{2(5)} = 48 \text{ MPa}$$

$$\sigma_\theta = 2\sigma_a = 96 \text{ MPa}$$

$$\tau_{xy} = \frac{-Tr}{J} = \frac{-10\pi(10^3)}{2\pi(0.12^2)(0.005)} = -69.4 \text{ MPa}$$

Hence,

$$\sigma_x = 48-25 = 23 \text{ MPa}$$
$$\sigma_y = 96 \text{ MPa}$$

Therefore,

$$\tau_{max} = \pm\left[\left(\frac{23-96}{2}\right)^2 + 69.4^2\right]^{1/2}$$

$$= \pm 78.4 \text{ MPa} \quad ◄$$

Also

$$\sigma' = \tfrac{1}{2}(23+96) = 59.5 \text{ MPa}$$

and

$$\theta_s = \tfrac{1}{2}\tan^{-1}\frac{23-96}{2(-69.4)} = -13.87°$$

Equation (1. 14b) with $\theta_s = -13.87°$:

$$\tau_{x'y'} = -16.99 - 61.42 = -78.4 \text{ MPa}$$

Thus,

$$\theta_s'' = 13.87° \quad \text{(Fig. b)} \quad ◄$$

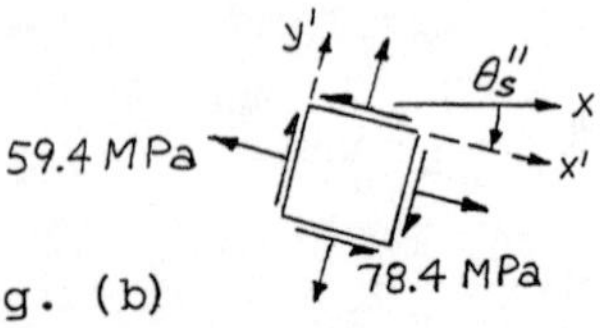

Fig. (b)

1.16

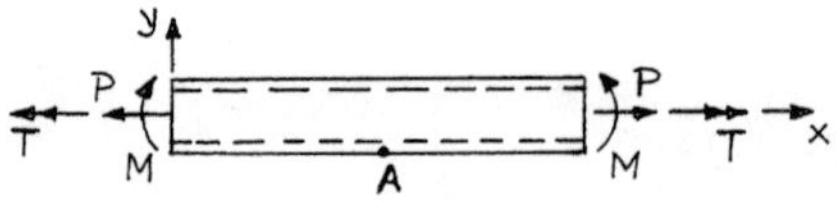

$A = \pi(30^2 - 15^2) = 2.121(10^{-3}) \text{ m}^2$

$I = \pi(30^4 - 15^4) = 0.596(10^{-6}) \text{ m}^4$

$J = 2I$

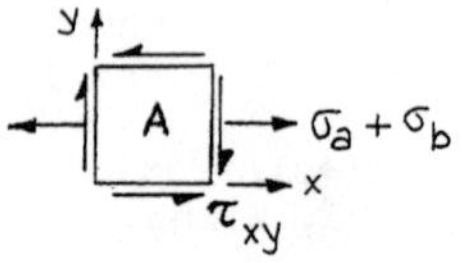

We have

$$\sigma_a = \frac{P}{A} = \frac{50(10^3)}{2.121(10^{-3})} = 23.58 \text{ MPa}$$

$$\sigma_b = \frac{Mr}{I} = \frac{200(0.03)}{0.596(10^{-6})} = 10.07 \text{ MPa}$$

$$\tau_{xy} = \frac{-Tr}{J} = \frac{-500(0.03)}{1.192(10^{-6})} = -12.58 \text{ MPa}$$

Thus,

$$\sigma_x = 23.58 + 10.07 = 33.65 \text{ MPa}$$
$$\sigma' = 16.83 \text{ MPa}$$

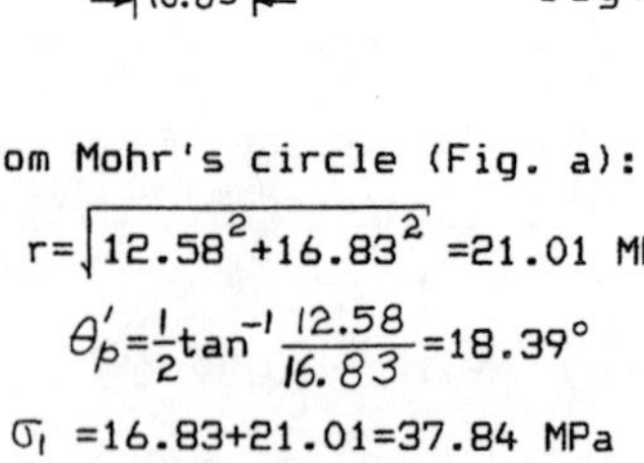

Fig. (a)

From Mohr's circle (Fig. a):

$$r = \sqrt{12.58^2 + 16.83^2} = 21.01 \text{ MPa}$$

$$\theta_p' = \tfrac{1}{2}\tan^{-1}\frac{12.58}{16.83} = 18.39°$$

$$\sigma_1 = 16.83 + 21.01 = 37.84 \text{ MPa}$$
$$\sigma_2 = -4.18 \text{ MPa} \quad ◄$$

Results are shown in Fig. (b).

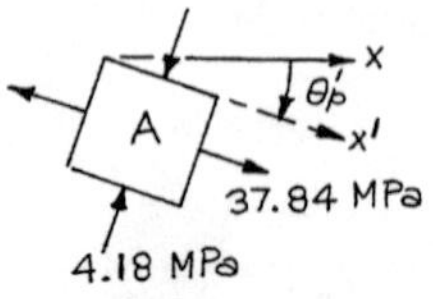

Fig. (b)

1.17

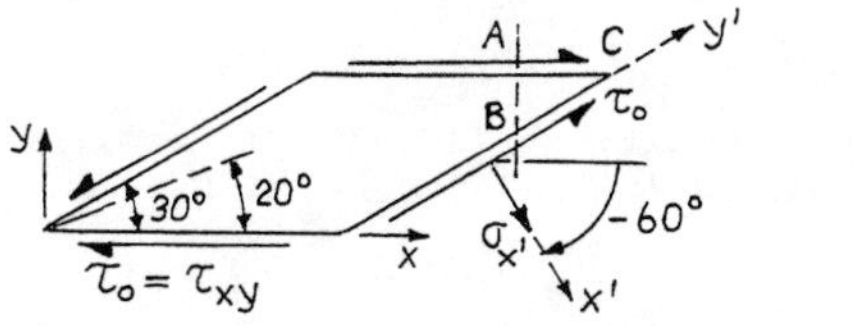

Fig. (a)

(a) At $\theta = -60°$ (Fig. a):

$$0 = \frac{\sigma_x + \sigma_y}{2} + \frac{\sigma_x - \sigma_y}{2}\cos 2(-60°) + \tau_o \sin 2(-60°)$$

or

$$0.5\sigma_x + 1.5\sigma_y - 1.738\tau_o \qquad \text{(a)}$$

We also have

$$\tau_o = -\frac{\sigma_x - \sigma_y}{2}\sin 2(-60°) + \tau_o \cos 2(-60°)$$

or

$$\sigma_x = 3.464\tau_o + \sigma_y \qquad \text{(b)}$$

Substituting Eq. (b) into (a), we obtain $\sigma_y = 0$. Results are shown in Fig. b.

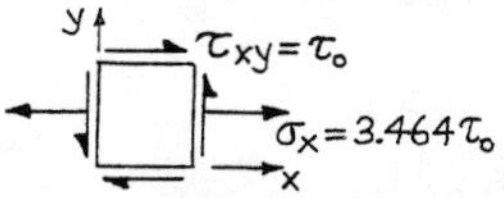

Fig. (b)

Alternatively, using an element ABC (Fig. c):

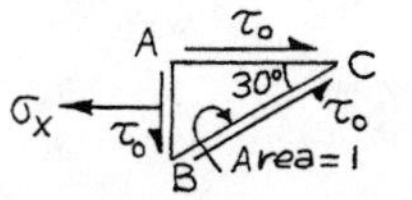

Fig. (c)

$$\Sigma F_x = 0.5\sigma_x - 0.866\tau_o - 0.866\tau_o = 0$$

or

$$\sigma_x = 3.464\tau_o \quad \text{(as before)}$$

Stresses on planes at 20°, taking $\theta = -70°$ (*Fig. b*):

$$\sigma_{20} = \left[\frac{3.464}{2} + \frac{3.464}{2}\cos(-140°) + \sin(-140°)\right]\tau_o$$

$$= -0.237\tau_o \qquad ◄$$

$$\tau_{20} = \left[-\frac{3.464}{2}\sin(-140°) + \cos(-140°)\right]\tau_o$$

$$= 0.347\tau_o \qquad ◄$$

(b) Principal stresses:

$$\sigma_{1,2} = \frac{3.464}{2} \pm \left[\left(\frac{3.464}{2}\right)^2 + \tau_o^2\right]^{1/2}$$

$$\sigma_1 = 3.732\tau_o \quad \sigma_2 = -0.268\tau_o \qquad ◄$$

The maximum principal stress is on plane inclined at

$$\theta_p' = \frac{1}{2}\tan^{-1}\frac{\tau_o}{1.732\tau_o} = 15° \qquad ◄$$

1.18

At a critical point on the shaft surface, the state of stress is as shown in Fig. (a).

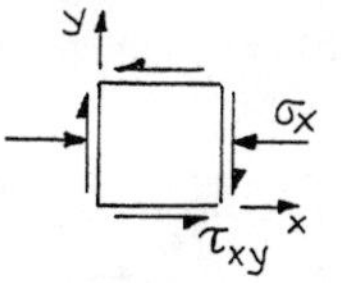

Fig. (a)

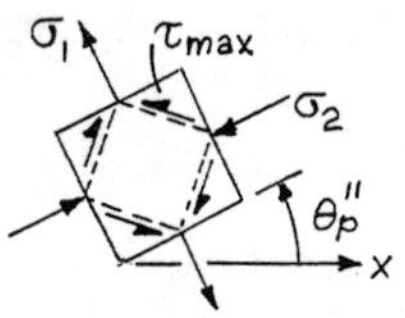

Fig. (b)

We have

$$\sigma_x = -\frac{P}{A} - \frac{Mr}{I}$$

$$= -\frac{81(10^3)}{\pi(0.075)^2} - \frac{13(10^3)(0.075)}{\pi(0.075)^4/4}$$

$$= -43.818 \text{ MPa}$$

$$\tau_{xy} = \frac{-Tr}{J} = -\frac{(15.6\times10^3)0.075}{\pi(0.075)^4/2} = -23.54 \text{ MPa}$$

Thus,

$$\sigma_{1,2} = \frac{-43.818}{2} \pm \left[\left(\frac{43.818}{2}\right)^2 + (-23.54)^2\right]^{1/2}$$

or

$$\sigma_1 = 10.248 \text{ MPa}, \quad \sigma_2 = -54.066 \text{ MPa}$$

$$\tau_{max} = \frac{1}{2}(\sigma_1 - \sigma_2) = 32.157 \text{ MPa} \qquad ◄$$

and

$$\theta_p'' = \frac{1}{2}\tan^{-1}\frac{2(23.54)}{43.818} = 23.53° \qquad ◄$$

Results are shown in Fig. b.

1.19

Apply Eqs. (1.14) to Fig.P1.19b, for $\theta = -30°$:

$$\sigma_{xb} = -40\sin 2(-30°) = 20\sqrt{3} \text{ MPa}$$
$$\sigma_{yb} = -20\sqrt{3} \text{ MPa} \qquad \text{(b)}$$
$$\tau_{xyb} = -40\cos 2(-30°) = -20 \text{ MPa}$$

Now apply Eqs. (1.14) to Fig. P1.19c, for $\theta = -60°$:

$$\sigma_{xc} = 10\sin 2(-60\ ) = -5\sqrt{3} \text{ MPa}$$
$$\sigma_{yc} = 5\sqrt{3} \text{ MPa} \qquad \text{(c)}$$
$$\tau_{xyc} = 10\cos 2(-60°) = -5 \text{ MPa}$$

Superposing stresses in Eqs. (b) and (c) and those in Fig. P1.19a, we obtain Fig. (a).

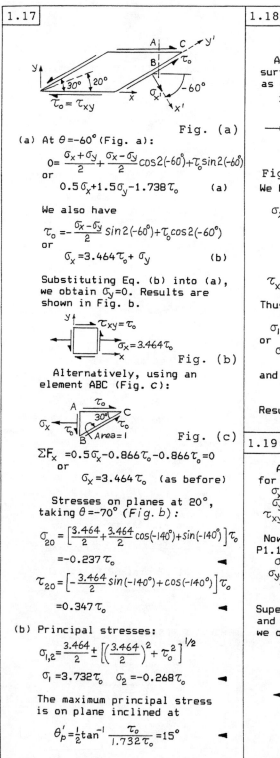

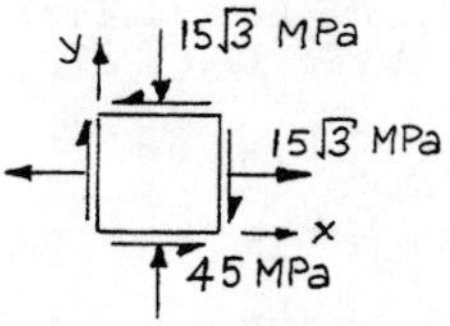

Fig. (a)

(CONT.)

1.19 CONT.

Referring to Fig. (a):

$$\sigma_{1,2}=0\pm[(15\sqrt{3})^2+(-45)^2]^{1/2}$$

or

$\sigma_1=51.96$ MPa, $\sigma_2=-51.96$ MPa ◄

When

$$\theta_p'=\frac{1}{2}\tan^{-1}\frac{2(-45)}{2(15\sqrt{3})}=-30^\circ$$ ◄

is substituted into Eq. (1.14a), we have 51.96 MPa (Fig. b).

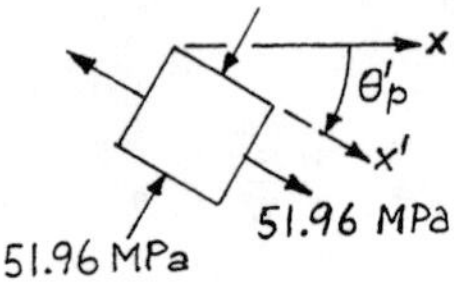

Fig. (b)

1.20

Apply Eqs. (1.14) to Fig. P1.20a, for $\theta=-15^\circ$, to obtain stresses in Fig. (a):

$$\sigma_{xa}=-\frac{30}{2}-\frac{30}{2}\cos 2(-15^\circ)=-27.99 \text{ MPa}$$

$$\sigma_{ya}=-15+15\cos 2(-15^\circ)=-2.01 \text{ MPa}$$

$$\tau_{xya}=15\sin 2(-15^\circ)=-7.5 \text{ MPa}$$

Superposition of stresses in Figs. (a) and P1.20b gives Fig. (b).

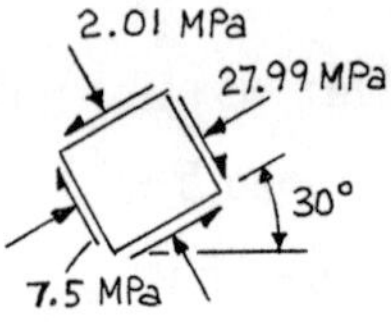

Fig. (a)

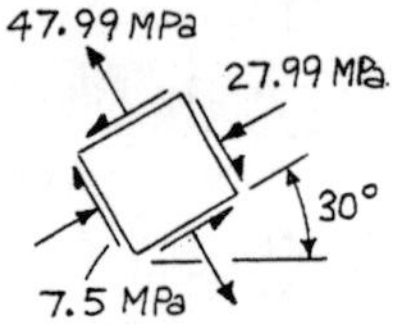

Fig. (b)

Apply Eq. (1.16) to Fig. (b):

$$\sigma_{1,2}=(-27.99+47.99)/2\pm[\tfrac{1}{4}(-27.99-47.99)^2+(-7.5)^2]^{1/2}$$

or

$\sigma_1=48.72$ MPa, $\sigma_2=-28.72$ MPa ◄

When

$$\theta_p=\frac{1}{2}\tan^{-1}\frac{2(-7.5)}{-(27.99+47.99)}=5.58^\circ$$

is substituted into Eq. (1.14a), we obtain -28.72 MPa (Fig. c).

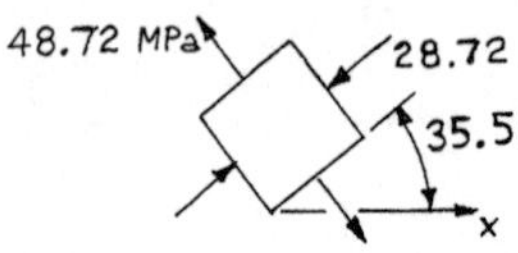

Fig. (c)

1.21

Equations (1.14) are applied to Fig. P1.21a, for $\theta=-30^\circ$:

$$\sigma_{xa}=\frac{20+30}{2}+\frac{20-30}{2}\cos 2(-30^\circ)$$

$$=22.5 \text{ MPa}$$

$$\sigma_{ya}=25-(-5)\cos 2(-30^\circ)$$

$$=27.5 \text{ MPa}$$

$$\tau_{xya}=-(-5)\sin 2(-30^\circ)=-4.33 \text{ MPa}$$

These stresses and that of Fig. P1.21b are superimposed to yield Fig. (a).

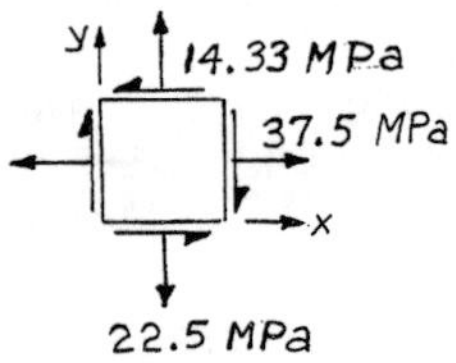

Fig. (a)

Principal stresses are thus

$$\sigma_{1,2}=\frac{37.5+22.5}{2}\pm\left[\left(\frac{37.5-22.5}{2}\right)^2+14.33^2\right]^{1/2}$$

or

$\sigma_1=46.17$ MPa $\quad\sigma_2=13.83$ MPa ◄

Hence

$$\tau_{max}=\frac{1}{2}(\sigma_1-\sigma_2)=16.17 \text{ MPa}$$ ◄

We have

$$\theta_p=\frac{1}{2}\tan^{-1}\frac{2(-14.33)}{37.5-22.5}=-31.2^\circ$$

Equation (1.14a) results in

$$\sigma_{x'}=\frac{37.5+22.5}{2}+\frac{37.5-22.5}{2}\cos(-62.4^\circ)$$

$$-14.33\sin(-62.4^\circ)=46.17 \text{ MPa}$$

Therefore

$$\theta_p'=31.2^\circ$$ ◄

Results are shown in a properly oriented element in Fig. (b).

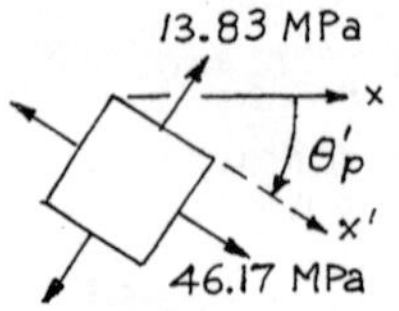

Fig. (b)

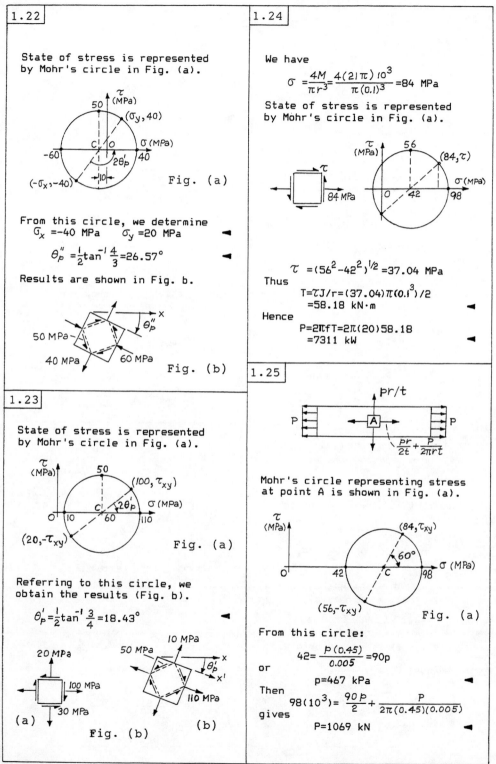

1.22

State of stress is represented by Mohr's circle in Fig. (a).

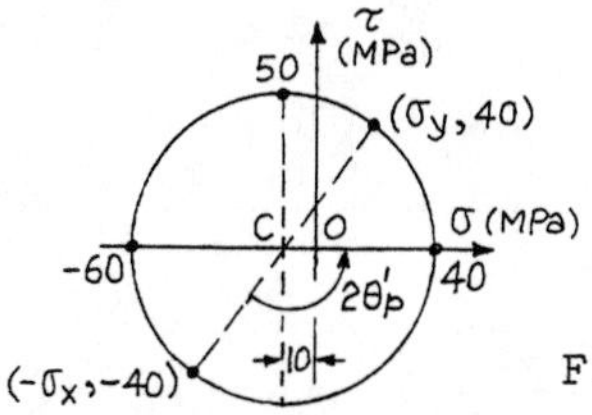

Fig. (a)

From this circle, we determine

$\sigma_x = -40$ MPa $\quad \sigma_y = 20$ MPa ◄

$$\theta_p'' = \frac{1}{2}\tan^{-1}\frac{4}{3} = 26.57°$$ ◄

Results are shown in Fig. b.

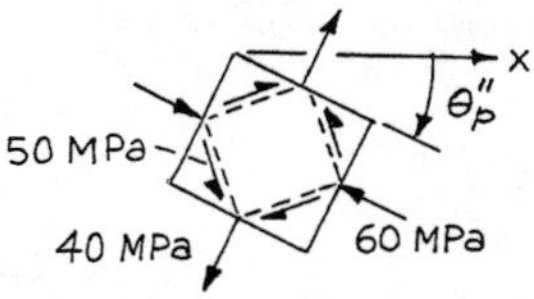

Fig. (b)

1.23

State of stress is represented by Mohr's circle in Fig. (a).

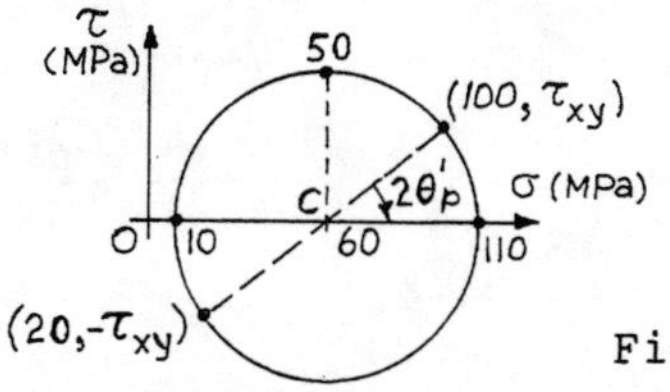

Fig. (a)

Referring to this circle, we obtain the results (Fig. b).

$$\theta_p' = \frac{1}{2}\tan^{-1}\frac{3}{4} = 18.43°$$ ◄

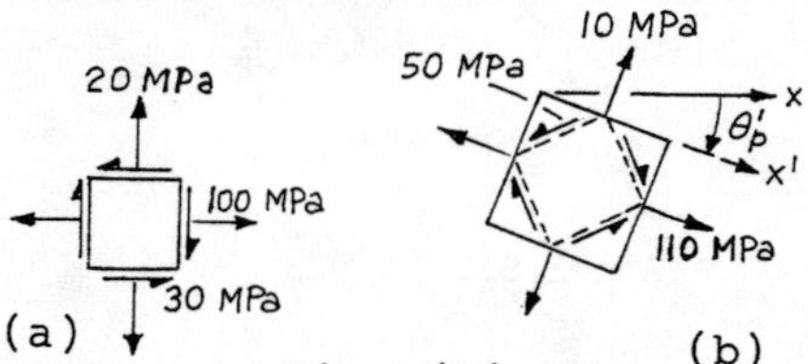

Fig. (b)

1.24

We have

$$\sigma = \frac{4M}{\pi r^3} = \frac{4(21\pi)10^3}{\pi(0.1)^3} = 84 \text{ MPa}$$

State of stress is represented by Mohr's circle in Fig. (a).

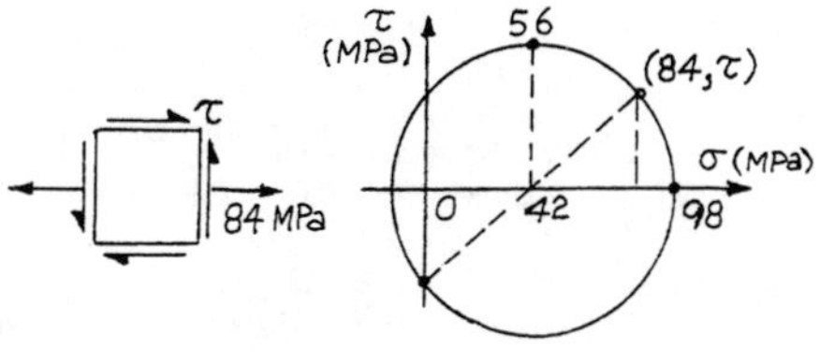

$$\tau = (56^2 - 42^2)^{1/2} = 37.04 \text{ MPa}$$

Thus

$$T = \tau J/r = (37.04)\pi(0.1^3)/2$$
$$= 58.18 \text{ kN·m}$$ ◄

Hence

$$P = 2\pi f T = 2\pi(20)58.18$$
$$= 7311 \text{ kW}$$ ◄

1.25

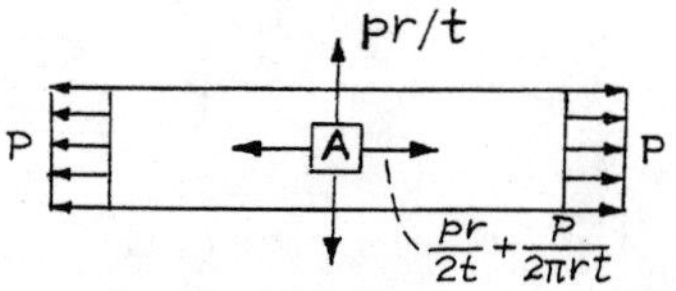

Mohr's circle representing stress at point A is shown in Fig. (a).

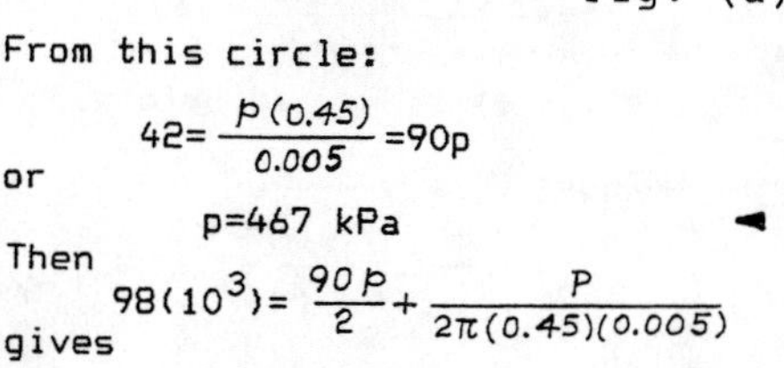

Fig. (a)

From this circle:

$$42 = \frac{p(0.45)}{0.005} = 90p$$

or

$$p = 467 \text{ kPa}$$ ◄

Then

$$98(10^3) = \frac{90p}{2} + \frac{P}{2\pi(0.45)(0.005)}$$

gives

$$P = 1069 \text{ kN}$$ ◄

1.26

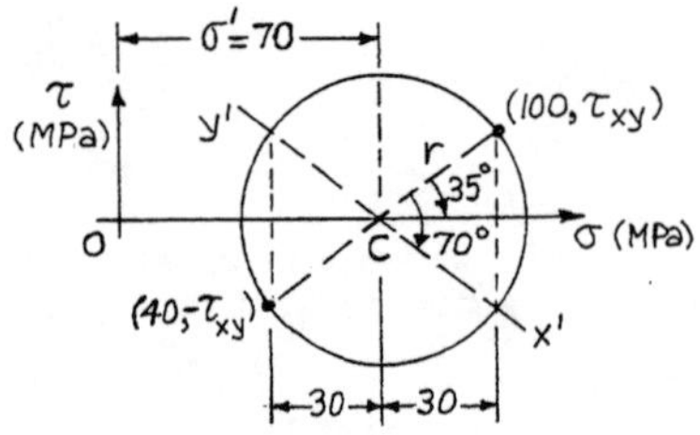

$$r=30/\cos 35^\circ=36.62\ \text{MPa}$$

(a) $\tau_{xy}=-36.62\sin 35^\circ=-21$ MPa ◀

(b) Because of symmetry:

$$\tau_{x'y'}=-\tau_{xy}=21\ \text{MPa}$$ ◀

and

$$\sigma_x+\sigma_y=\sigma_{x'}+\sigma_{y'}=140\ \text{MPa}$$

gives

$$\sigma_{y'}=40\ \text{MPa}$$ ◀

1.27

State of stress is represented by Mohr's circle in Fig. (a).

Fig. (a)

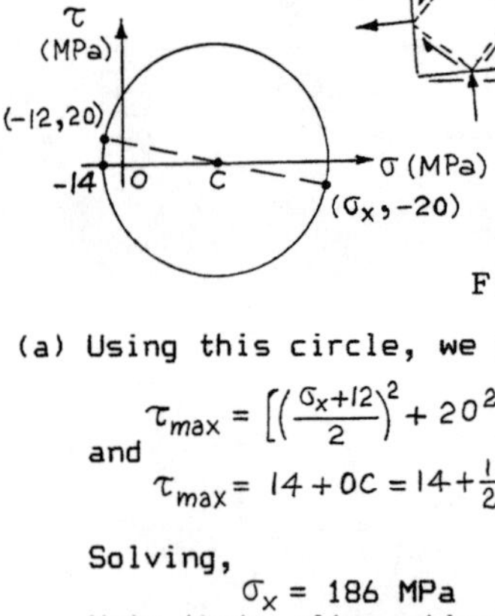

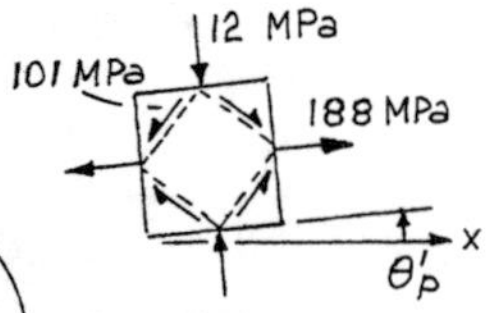

Fig. (b)

(a) Using this circle, we write

$$\tau_{max}=\left[\left(\frac{\sigma_x+12}{2}\right)^2+20^2\right]^{1/2}$$

and

$$\tau_{max}=14+OC=14+\frac{1}{2}(\sigma_x-12)$$

Solving,

$$\sigma_x=186\ \text{MPa}$$ ◀

Note that, alternatively,

$$-14=\frac{\sigma_x-12}{2}-\left[\left(\frac{\sigma_x+12}{2}\right)^2+20^2\right]^{1/2}$$

yields σ_x=186 MPa, as before.

(b) We have

$$\sigma_{1,2}=\frac{186-12}{2}\pm\left[\left(\frac{186+12}{2}\right)^2+20^2\right]$$

or

$\sigma_1=188$ MPa $\sigma_2=-14$ MPa ◀

(CONT.)

1.27 CONT.

and

$$\tau_{max}=\frac{1}{2}(\sigma_1-\sigma_2)=101\ \text{MPa}$$ ◀

Also

$$\theta_p'=\frac{1}{2}\tan^{-1}\frac{2(20)}{186+12}=5.68^\circ$$ ◀

Results are shown in Fig. (b).

1.28

(a) $\sigma_1=96.05$ MPa $\sigma_2=23.05$ MPa $\sigma_3=0$

(b) $(\tau_{12})_{max}=\frac{1}{2}(\sigma_1-\sigma_2)=36.05$ MPa

$(\tau_{13})_{max}=\frac{1}{2}(\sigma_1-\sigma_3)=48.03$ MPa ◀

$(\tau_{23})_{max}=\frac{1}{2}(\sigma_2-\sigma_3)=11.98$ MPa

Plane of $(\tau_{12})_{max}$ is shown in Fig. (a). Other maximum shear planes are sketched similarly.

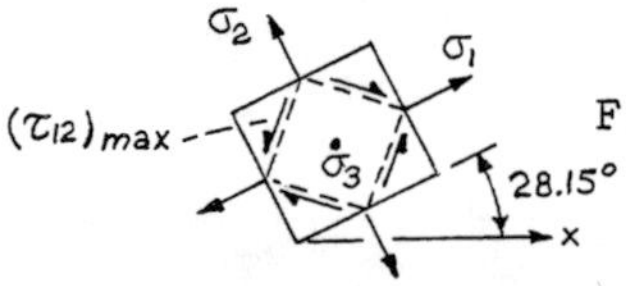

Fig. (a)

1.29

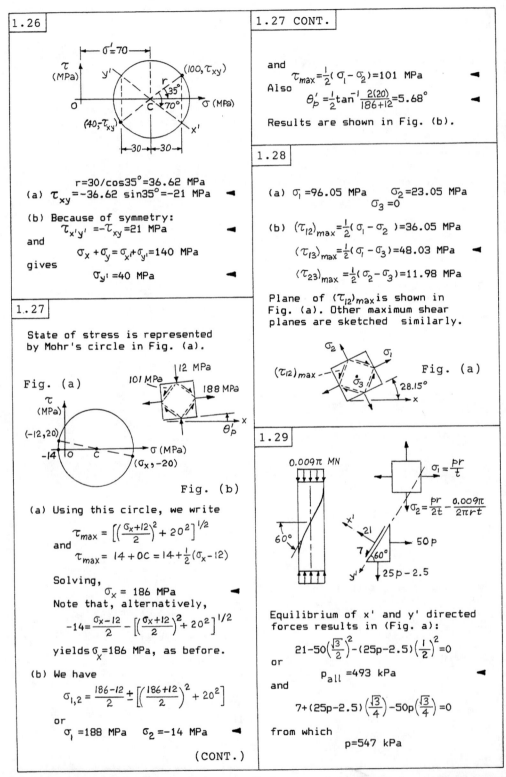

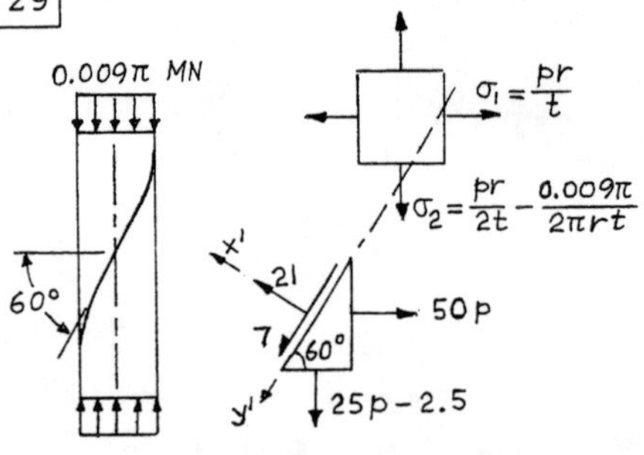

Equilibrium of x' and y' directed forces results in (Fig. a):

$$21-50\left(\frac{\sqrt{3}}{2}\right)^2-(25p-2.5)\left(\frac{1}{2}\right)^2=0$$

or

$$p_{all}=493\ \text{kPa}$$ ◀

and

$$7+(25p-2.5)\left(\frac{\sqrt{3}}{4}\right)-50p\left(\frac{\sqrt{3}}{4}\right)=0$$

from which

$$p=547\ \text{kPa}$$

1.30

Direction cosines are:

$l_1=\sqrt{3}/2$ $m_1=1/2$ $n_1=0$
$l_2=-1/2$ $m_2=\sqrt{3}/2$ $n_2=0$
$l_3=0$ $m_3=0$ $n_3=1$

Equations (1.24a) is thus

$\sigma_{x'}=20(3/4)+0+0+2(12)(\sqrt{3}/2)(1/2)$
$+0+0=25.392$ MPa

Similarly, applying Eqs. (1.24b) through (1.24e), we obtain $[\tau_{i'j'}]$:

$$\begin{bmatrix} 25.342 & -2.66 & -7.99 \\ -2.66 & -5.392 & 16.16 \\ -7.99 & 16.16 & 6 \end{bmatrix} \text{MPa} \quad ◄$$

Then, Eqs. (1.29) result in

$I_1=I_1'=26$ MPa

$I_2=I_2'=-349\ (\text{MPa})^2$ ◄

$I_3=I_3'=-6464\ (\text{MPa})^3$

1.31

Direction cosines are:

$l_1=\sqrt{3}/2$ $m_1=1/2$ $n_1=0$
$l_2=-1/2$ $m_2=\sqrt{3}/2$ $n_2=0$
$l_3=0$ $m_3=0$ $n_3=1$

Equation (1.24a) is therefore

$\sigma_{x'}=60(3/4)+0+0+2(40)(\sqrt{3}/2)(1/2)$
$+0+0=20[(9/4)+\sqrt{3}]=79.64$ MPa

Similarly, applying Eqs. (1.24b) through (1.24e), we obtain $[\tau_{ij}]$:

$$\begin{bmatrix} 79.64 & -5.98 & -44.64 \\ -5.98 & -19.64 & -2.68 \\ -44.64 & -2.68 & 20 \end{bmatrix} \text{MPa} \quad ◄$$

Then, Eqs. (1.29) lead to

$I_1=I_1'=80$ MPa

$I_2=I_2'=-2400\ (\text{MPa})^2$ ◄

$I_3=I_3'=8000\ (\text{MPa})^3$

1.32

Referring to Appendix B:

$\sigma_1=13.212$ MPa $\sigma_2=5.684$ MPa
$\sigma_3=-8.896$ MPa

and

$l_1=0.9556$ $m_1=0.1688$
$n_1=0.2416$

Thus,

$\tau_{max}=\frac{1}{2}(\sigma_1-\sigma_3)=11.054$ MPa ◄

1.33

Referring to Appendix B:

$\sigma_1=66.061$ MPa $\sigma_2=28.418$ MPa
$\sigma_3=-44.479$ MPa ◄

and

$l_1=0.9556$ $m_1=0.1688$
$n_1=0.2416$

1.34

Referring to Appendix B:

$\sigma_1=30.493$ MPa $\sigma_2=12.485$ MPa
$\sigma_3=-16.979$ MPa

Thus,

$\tau_{max}=\frac{1}{2}(\sigma_1-\sigma_3)=23.736$ MPa ◄

1.35

Referring to Appendix B:

$\sigma_1=24.747$ MPa $\sigma_2=8.480$ MPa ◄
$\sigma_3=2.773$ MPa

and

$l_1=0.6467$, $m_1=0.3958$, $n_1=0.6421$

1.36

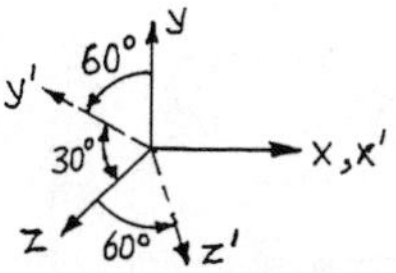

Fig. (a)

(a) At point (3,1,5) with respect to xyz axes, we have $[\tau_{ij}]$:

$$\begin{bmatrix} 10 & 0 & 0 \\ 0 & -4 & 0 \\ 0 & 0 & 8 \end{bmatrix} \text{MPa} \qquad \text{(a)}$$

Then, Eqs. (1.29) result in

$I_1=14$ MPa
$I_2=8\ (\text{MPa})^2$ ◄
$I_3=-320\ (\text{MPa})^3$

Direction cosines of x'y'z', referring to Fig. (a) are

$l_1=1$ $m_1=0$ $n_1=0$
$l_2=0$ $m_2=1/2$ $n_2=\sqrt{3}/2$
$l_3=0$ $m_3=-\sqrt{3}/2$ $n_3=1/2$

Now Eqs. (1.24) and (a) give $[\tau_{i'j'}]$:

$$\begin{bmatrix} 10 & 0 & 0 \\ 0 & 5 & 3\sqrt{3} \\ 0 & 3\sqrt{3} & -1 \end{bmatrix} \text{MPa} \quad ◄$$

Thus, Eqs. (1.29) yield

$I_1'=14$ MPa $I_2'=8\ (\text{MPa})^2$
$I_3'=-320\ (\text{MPa})^3$ ◄

as before.

(CONT.)

1.36 CONT.

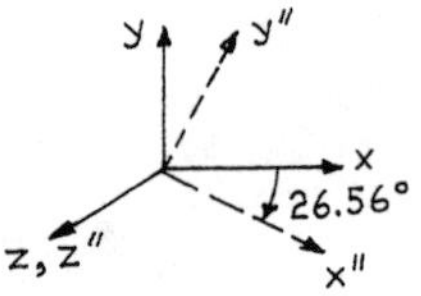

Fig. (b)

(b) Direction cosines are (Fig.b):

$l_1=2/\sqrt{5}$ $m_1=-1/\sqrt{5}$ $n_1=0$

$l_2=1/\sqrt{5}$ $m_2=2/\sqrt{5}$ $n_2=0$

$l_3=0$ $m_3=0$ $n_3=1$

With these and Eq. (a), Eqs. (1.24) yield $[\tau_{i'j'}]$:

$$\begin{bmatrix} 7.2 & 5.6 & 0 \\ 5.6 & -1.2 & 0 \\ 0 & 0 & 8 \end{bmatrix} \text{MPa}$$ ◄

Then Eqs. (1.29) result in

$I_1''=14$ MPa $I_2''=8$ $(\text{MPa})^2$ ◄
$I_3''=-320$ $(\text{MPa})^3$

The I's are thus invariants.

1.37

Introducing the given data into Eq. (1.24a), we obtain

$$\sigma_{x'}=12(1/2)^2+10(\sqrt{3}/2)^2+14(0)$$
$$+2[6(1/2)(\sqrt{3}/2)+0+0$$
$$=15.696 \text{ MPa}$$

Remaining stress components are determined in a like manner. The result, $[\tau_{i'j'}]$, is

$$\begin{bmatrix} 15.696 & -3.866 & 7.098 \\ -3.866 & 6.304 & -6.294 \\ 7.098 & -6.294 & 14 \end{bmatrix} \text{MPa}$$ ◄

1.38

Equations (1.29) become

$$I_1=\sigma_x+\sigma_y \qquad I_2=\sigma_x\cdot\sigma_y-\tau_{xy}^2$$
$$I_3=0$$

Equation (1.28) is then

$$\sigma_p^3-(\sigma_x+\sigma_y)\sigma_p^2+(\sigma_x\sigma_y-\tau_{xy}^2)\sigma_p=0$$

or

$$\sigma_p^2-(\sigma_x+\sigma_y)\sigma_p+(\sigma_x\sigma_y-\tau_{xy}^2)=0$$

Solution of this quadratic equation is

$$\sigma_{1,2}=\frac{\sigma_x+\sigma_y}{2}\pm\frac{1}{2}\left[\sigma_x^2+2\sigma_x\sigma_y+\sigma_y^2-4(\sigma_x+\sigma_y-\tau_{xy}^2)\right]^{1/2}$$

$$=\frac{\sigma_x+\sigma_y}{2}\pm\left[\left(\frac{\sigma_x-\sigma_y}{2}\right)^2+\tau_{xy}^2\right]^{1/2}$$ ◄

1.39

Referring to Appendix B, we obtain the following values.

(a) $\sigma_1=12.049$ MPa
$\sigma_2=-1.521$ MPa ◄
$\sigma_3=-4.528$ MPa

and $l_1=0.6184$
$m_1=0.5333$
$n_1=0.5772$

(b) $\sigma_1=19.238$ MPa
$\sigma_2=13.704$ MPa ◄
$\sigma_3=4.648$ MPa

and $l_1=0.3339$
$m_1=0.3862$
$n_1=0.8599$

1.40

(a) Direction cosines are:

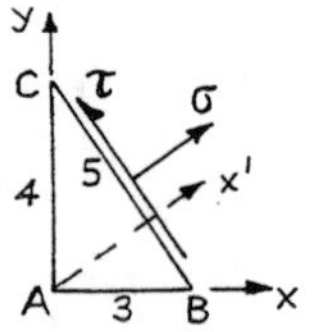

$l=4/5$
$=0.8$
$m=3/5$
$=0.6$
$n=0$

Equation (1.35) is thus

$$\sigma=100(0.8)^2+60(0.6)^2+2(40)(0.8)(0.6)$$
$$=124 \text{ MPa}$$ ◄

Equation (1.22) yield

$$p_x=100(0.8)+40(0.6)$$
$$=104 \text{ MPa}$$
$$p_y=40(0.8)+60(0.6)=68 \text{ MPa}$$
$$p_z=80(0.6)=48 \text{ MPa}$$

Equation (1.36) is then

$$\tau=[104^2+68^2+48^2-124^2]^{1/2}$$
$$=48.66 \text{ MPa}$$ ◄

(b) Direction cosines are:

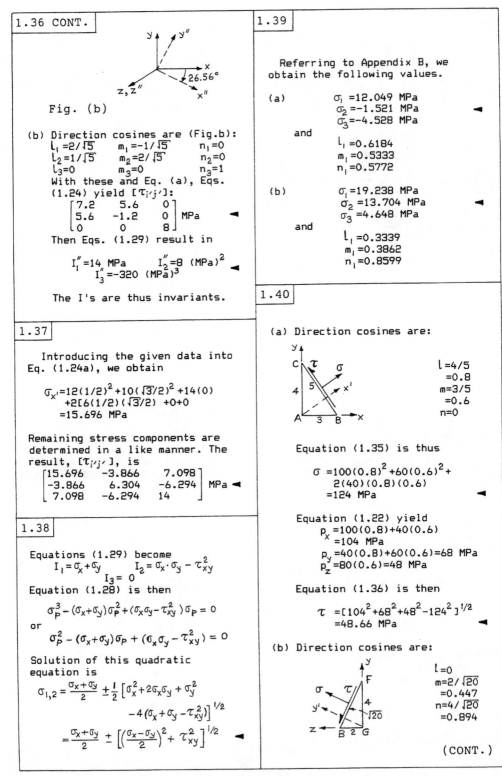

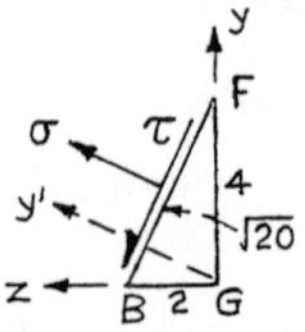

$l=0$
$m=2/\sqrt{20}$
$=0.447$
$n=4/\sqrt{20}$
$=0.894$

(CONT.)

1.40 CONT.

Equation (1.35) is thus

$$\sigma = 60(0.447)^2 + 20(0.894)^2 + 2(80)(0.447)(0.894) = 91.913 \text{ MPa}$$ ◄

Equations (1.22) give

$p_x = 40(0.447) = 17.88$ MPa

$p_y = 60(0.447) + 80(0.894) = 98.34$ MPa

$p_z = 80(0.447) + 20(0.894) = 53.64$ MPa

Equation (1.36) yields then

$$\tau = [17.88^2 + 98.34^2 + 53.66^2 - 91.913^2]^{1/2} = 66.481 \text{ MPa}$$ ◄

(c) Direction cosines are:

$l = 0.512 \quad m = 0.384 \quad n = 0.768$

Equation (1.35) is therefore

$$\sigma = 100(0.512)^2 + 60(0.384)^2 + 20(0.768)^2 + 2[40(0.512)(0.384) + 80(0.384)(0.768)] = 109.77 \text{ MPa}$$ ◄

Equations (1.22) yield

$p_x = 100(0.512) + 40(0.384) = 66.56$ MPa

$p_y = 40(0.512) + 60(0.384) + 80(0.768) = 104.96$ MPa

$p_z = 80(0.384) + 20(0.768) = 46.08$ MPa

Equation (1.36) gives

$$\tau = [66.56^2 + 104.96^2 + 46.08^2 - 109.77^2]^{1/2} = 74.3 \text{ MPa}$$ ◄

1.41

(a) Direction cosines are:

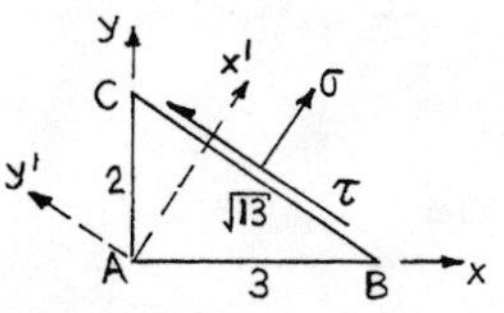

$l = 2/\sqrt{13} = 0.555$

$m = 3/\sqrt{13} = 0.882$

$n = 0$

(CONT.)

1.41 CONT.

Equation (1.35) is then

$$\sigma = 100(0.555)^2 + 60(0.832)^2 + 2(40)(0.555)(0.832) = 109.277 \text{ MPa}$$ ◄

Equations (1.22) lead to

$p_x = 100(0.555) + 40(0.832) = 88.78$ MPa

$p_y = 40(0.555) + 60(0.832) = 72.12$ MPa

$p_z = 80(0.832) = 66.56$ MPa

Equation (1.36) gives then

$$\tau = [88.78^2 + 72.12^2 + 66.56^2 - 109.277^2]^{1/2} = 74.67 \text{ MPa}$$ ◄

(b) Direction cosines are:

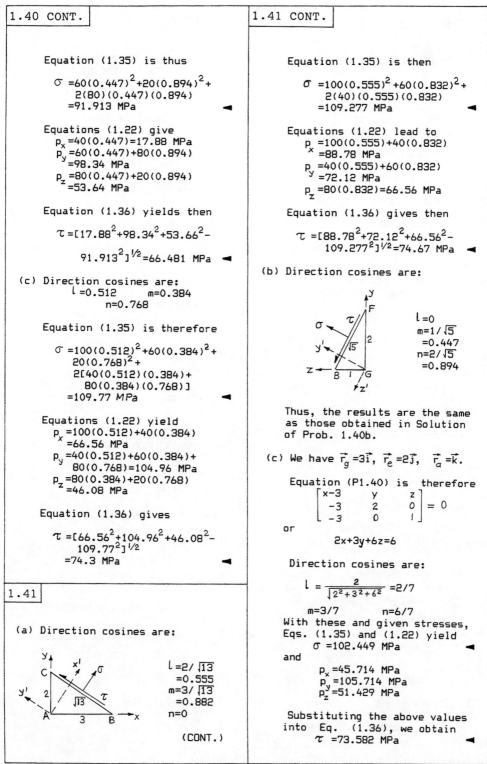

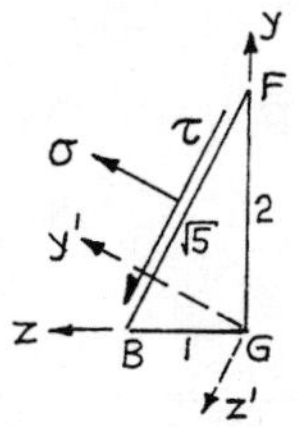

$l = 0$

$m = 1/\sqrt{5} = 0.447$

$n = 2/\sqrt{5} = 0.894$

Thus, the results are the same as those obtained in Solution of Prob. 1.40b.

(c) We have $\vec{r}_g = 3\vec{i}$, $\vec{r}_e = 2\vec{j}$, $\vec{r}_a = \vec{k}$.

Equation (P1.40) is therefore

$$\begin{bmatrix} x-3 & y & z \\ -3 & 2 & 0 \\ -3 & 0 & 1 \end{bmatrix} = 0$$

or

$$2x + 3y + 6z = 6$$

Direction cosines are:

$$l = \frac{2}{\sqrt{2^2+3^2+6^2}} = 2/7$$

$m = 3/7 \qquad n = 6/7$

With these and given stresses, Eqs. (1.35) and (1.22) yield

$\sigma = 102.449$ MPa ◄

and

$p_x = 45.714$ MPa

$p_y = 105.714$ MPa

$p_z = 51.429$ MPa

Substituting the above values into Eq. (1.36), we obtain

$\tau = 73.582$ MPa ◄

1.42

See: Hint, Prob. 1.40:

$$l=\frac{2}{\sqrt{3^2+1^2+2^2}}=\frac{2}{\sqrt{14}}$$

$$m=\frac{1}{\sqrt{14}} \qquad n=-\frac{3}{\sqrt{14}}$$

Equation (1.35) gives

$$\sigma=20(2/\sqrt{14})^2+30(1/\sqrt{14})^2+50(-3/\sqrt{14})^2+2[10(2/\sqrt{14})(1/\sqrt{14})-10(2/\sqrt{14})(-3/\sqrt{14})]$$
$$=51.43 \text{ MPa}$$ ◄

Equation (1.36):

$$\tau=\{[20(2/\sqrt{14})+10(1/\sqrt{14})-10(-3/\sqrt{14})]^2+[10(2/\sqrt{14})+30(1/\sqrt{14})+0]^2+[-10(2/\sqrt{14})+0+50(-3/\sqrt{14})]^2-51.43^2\}^{1/2}$$
$$=7.413 \text{ MPa}$$ ◄

1.43

Direction cosines are

$$l=\cos 35^\circ=0.8192$$
$$m=\cos 60^\circ=0.5$$
$$n=\cos 73.6^\circ=0.2823$$

Equation (1.35) gives

$$\sigma=60(0.8192)^2-40(0.5)^2+30(0.2823)^2+2[20(0.8192)(0.5)-5(0.5)(0.2823)+10(0.8192)(0.2823)]$$
$$=52.25 \text{ MPa}$$ ◄

Equations (1.22):

$$p_x=60(0.8192)+20(0.5)+10(0.2823)=61.9725 \text{ MPa}$$

and

(CONT.)

1.43 CONT.

$$p_y=20(0.8192)-40(0.5)-5(0.2823)=-5.0287 \text{ MPa}$$
$$p_z=10(0.8192)-5(0.5)+30(0.2823)=14.1618 \text{ MPa}$$

Equation (1.36) is thus

$$\tau=[(61.9725)^2+(-5.0287)^2+(14.1618)^2-(52.25)^2]^{1/2}$$
$$=36.56 \text{ MPa}$$ ◄

1.44

Direction cosines are

$$l=\cos 40^\circ=0.766$$
$$m=\cos 75^\circ=0.259$$
$$n=\cos 54^\circ=0.588$$

Equation (1.35):

$$\sigma=40(0.766)^2+20(0.259)^2+20(0.588)^2+2[40(0.766)(0.259)+0+30(0.766)(0.588)]$$
$$=23.47+1.34+6.91+42.9$$
$$=74.62 \text{ MPa}$$ ◄

Equation (1.36) gives

$$\tau=\{[40(0.766)+40(0.259)+30(0.588)]^2+[40(0.766)+20(0.259)+0]^2+[30(0.766)+0+20(0.588)]^2-74.62^2\}^{1/2}$$
$$=[3436.3+1282.9+1206.7-5568.1]^{1/2}$$
$$=18.93 \text{ MPa}$$ ◄

1.45

Note: Planes of maximum shear stresses can be determined upon following a procedure similar to that used in Solution of Prob. 1.28.

(a) From Problem 1.39a:

$\sigma_1 = 12.049$ MPa $\quad \sigma_2 = -1.521$ MPa $\quad \sigma_3 = -4.528$ MPa

Thus,

$$(\tau_{13})_{max} = \frac{1}{2}(\sigma_1 - \sigma_3) = 8.288 \text{ MPa}$$

$$(\tau_{12})_{max} = \frac{1}{2}(\sigma_1 - \sigma_2) = 6.785 \text{ MPa} \quad ◄$$

$$(\tau_{23})_{max} = \frac{1}{2}(\sigma_2 - \sigma_3) = 1.503 \text{ MPa}$$

(b) From Problem 1.39b:

$\sigma_1 = 19.237$ MPa $\quad \sigma_2 = 13.704$ MPa $\quad \sigma_3 = 4.648$ MPa

Thus,

$$(\tau_{13})_{max} = \frac{1}{2}(\sigma_1 - \sigma_3) = 7.294 \text{ MPa}$$

$$(\tau_{12})_{max} = \frac{1}{2}(\sigma_1 - \sigma_2) = 2.766 \text{ MPa} \quad ◄$$

$$(\tau_{23})_{max} = \frac{1}{2}(\sigma_2 - \sigma_3) = 4.528 \text{ MPa}$$

1.46

Octahedral and shearing stresses are given by

$$\tau_{oct}^2 = \frac{1}{9}[(\sigma_1 - \sigma_2)^2 + (\sigma_2 - \sigma_3)^2 + (\sigma_3 - \sigma_1)^2]$$

$$\tau_{max}^2 = \frac{1}{4}(\sigma_1 - \sigma_3)^2$$

Let us say, $\tau_{max}^2 > \tau_{oct}^2$. Then

$$\left(\frac{\sigma_1 - \sigma_3}{2}\right)^2 > \frac{1}{9}\left[(\sigma_1-\sigma_2)^2 + (\sigma_2-\sigma_3)^2 + (\sigma_3-\sigma_1)^2\right]$$

or

$$\frac{9}{4}(\sigma_1 - \sigma_3)^2 > [(\sigma_1 - \sigma_3)^2 + (\sigma_2 - \sigma_3)^2 + (\sigma_3 - \sigma_1)^2]$$

Subtracting $(\sigma_1 - \sigma_3)^2$ from both sides and noting that

$$(\sigma_1 - \sigma_3)^2 = (\sigma_3 - \sigma_1)^2$$

We have

$$\frac{5}{4}(\sigma_1 - \sigma_3)^2 > (\sigma_1 - \sigma_3)^2 + (\sigma_2 - \sigma_3)^2$$

(CONT.)

1.46 CONT.

But

$$(\sigma_1 - \sigma_3)^2 > (\sigma_1 - \sigma_2)^2 + (\sigma_2 - \sigma_3)^2$$

Thus,

$$\frac{5}{4}(\sigma_1 - \sigma_3)^2 > (\sigma_1 - \sigma_2)^2 + (\sigma_2 - \sigma_3)^2 \quad \text{(a)}$$

The squares of the difference between σ_1 and σ_3 will always be greater than the sum of the squares of the difference between σ_1 and σ_2, σ_2 and σ_3, since $\sigma_1 > \sigma_2 > \sigma_3$. Hence, Eq. (a) is true and our assumption is correct. That is

$$\tau_{max} > \tau_{oct} \quad ◄$$

1.47

From Solution of Problem 1.35:

$\sigma_1 = 24.747$ MPa $\quad \sigma_2 = 8.48$ MPa $\quad \sigma_3 = 2.773$ MPa

Applying Eqs. (1.38) and (1.39):

$$\tau_{oct} = \frac{1}{3}[(24.747-8.48)^2 + (8.48-2.773)^2 + (2.773-24.747)^2]^{1/2}$$

$$= 9.31 \text{ MPa} \quad ◄$$

and

$$\sigma_{oct} = \frac{1}{3}(24.747+8.48+2.773)$$

$$= 12 \text{ MPa} \quad ◄$$

Therefore

$$p_{oct} = (9.31^2 + 12^2)^{1/2} = 15.19 \text{ MPa} \quad ◄$$

1.48

Shearing stress, in terms of principal stresses, is given by

$$\tau^2 = \sigma_1^2 l^2 + \sigma_2 m^2 + \sigma_3^2 n^2 - (\sigma_1 l^2 + \sigma_2 m^2 + \sigma_3 n^2)^2 \quad \text{(a)}$$

We substitute $n^2 = 1 - m^2 - l^2$ into Eq. (a), calculate its derivatives with respect to l and m, and equate these derivatives to zero:

(CONT.)

1.48 CONT.

$$\frac{\partial\tau}{\partial l} = l\left[(\sigma_1-\sigma_3)l^2+(\sigma_2-\sigma_3)m^2-\frac{1}{2}(\sigma_3-\sigma_1)\right]=0 \quad (b)$$

$$\frac{\partial\tau}{\partial m} = m\left[(\sigma_1-\sigma_3)l^2+(\sigma_2-\sigma_3)m^2-\frac{1}{2}(\sigma_2-\sigma_3)\right]=0 \quad (c)$$

One solution is $l=m=0$. Solutions for the direction cosines of planes for which τ is a maximum or minimum can also be found as follows.

Take $l=0$: Eq. (c) gives $m=\pm\sqrt{1/2}$

Take $m=0$: Eq. (b) gives $l=\pm\sqrt{1/2}$

There are, in general, no solutions of Eqs. (b) and (c) in which l and m are both different from zero, for this case the expressions in brackets cannot both vanish.

By the above procedure we can form the following table.

Direction cosines for planes of τ_{max} and τ_{min}

$l=$	0	0	± 1	0	$\pm\sqrt{1/2}$	$\pm\sqrt{1/2}$
$m=$	0	± 1	0	$\pm\sqrt{1/2}$	0	$\pm\sqrt{1/2}$
$n=$	± 1	0	0	$\pm\sqrt{1/2}$	$\pm\sqrt{1/2}$	0

The first three columns define the planes for τ_{min}, where $\tau=0$. The last three columns give planes through each principal axes bisecting the angles between the two other principal axes. Substituting the latter direction cosines into Eq. (a), we have

$$(\tau_{23})_{max}=\pm\frac{\sigma_2-\sigma_3}{2} \qquad (\tau_{13})_{max}=\pm\frac{\sigma_1-\sigma_3}{2}$$

$$(\tau_{12})_{max}=\pm\frac{\sigma_1-\sigma_2}{2}$$ ◄

Similarly, introducing the direction cosines given in the above table into Eq. (1.32), we obtain the normal stresses associated with the maximum shearing stresses:

$$\sigma'_{12}=\frac{\sigma_1+\sigma_2}{2} \qquad \sigma'_{13}=\frac{\sigma_1+\sigma_3}{2}$$

$$\sigma'_{23}=\frac{\sigma_2+\sigma_3}{2}$$ ◄

1.49

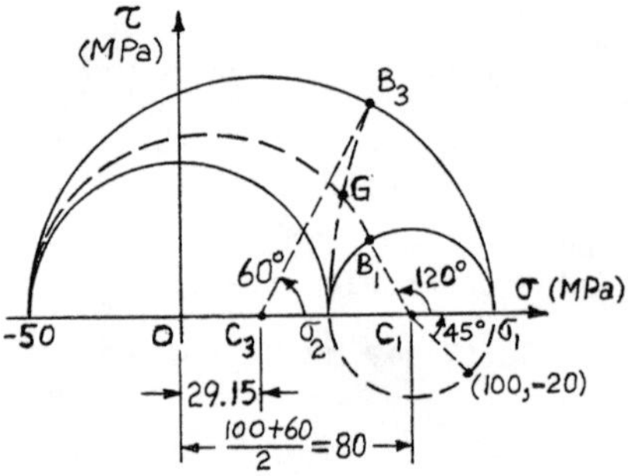

From Mohr's circle, we have

(a) $\sigma_1=108.3$ MPa $\quad \sigma_2=51.7$ MPa ◄

$\sigma_3=-50$ MPa $\quad \theta'_p=22.5°$

(b) $\tau_G=38$ MPa $\quad \sigma_G=56$ MPa ◄

1.50

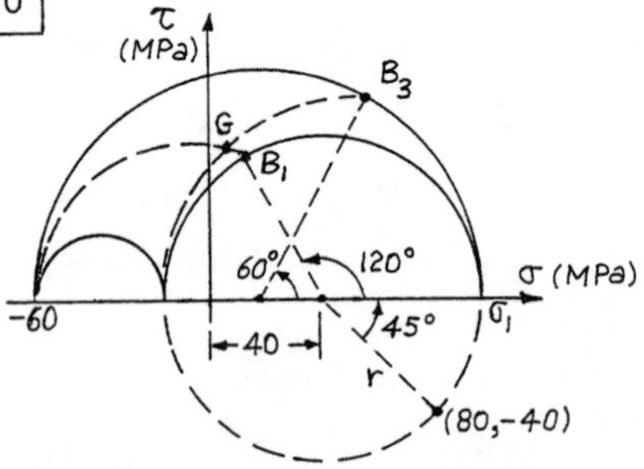

Referring to Mohr's circle:
$r=56.57$ MPa $\quad \theta'_p=22.5°$

$\sigma_1=96.57$ MPa $\quad \sigma_2=16.57$ MPa ◄
$\tau_G=50.7$ MPa $\quad \sigma_G=6.1$ MPa

1.51

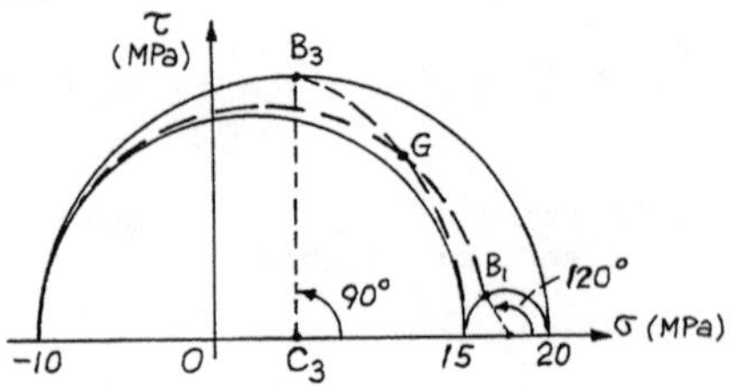

From Mohr's circle, we have

$\sigma_G=10.7$ MPa $\quad \tau_G=11$ MPa ◄

1.52

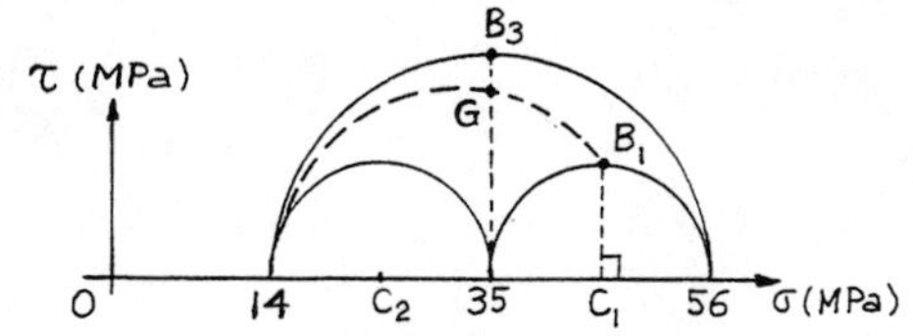

From Mohr's circle, we have

τ_{max}=21 MPa σ_{oct}=35 MPa

τ_{oct}=17.15 MPa ◄

1.53

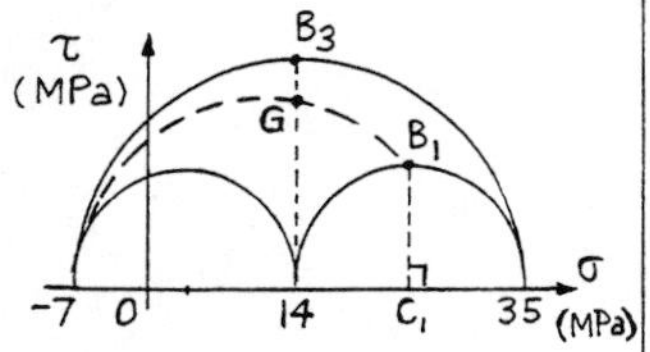

Referring to Mohr's circle:

τ_{max} =21 MPa σ_{oct}=14 MPa

τ_{oct}=17.15 MPa ◄

1.54

Approach I. Graphical Solution

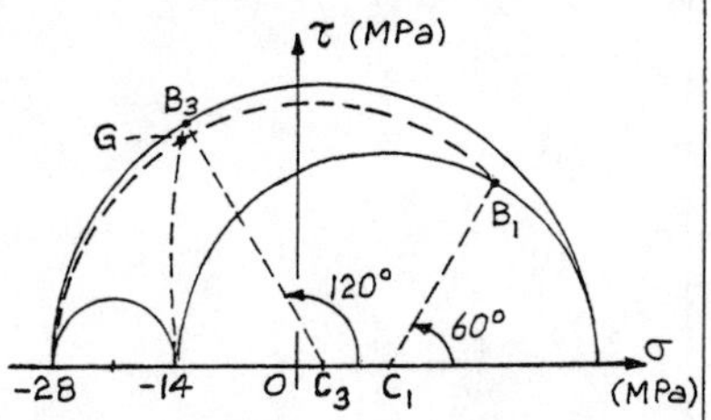

From Mohr's circle, we obtain

σ_G =-12.4 MPa τ_G =26.2 MPa ◄

Approach II. Analytical Solution

Direction cosines of the oblique n-plane are found as follows.

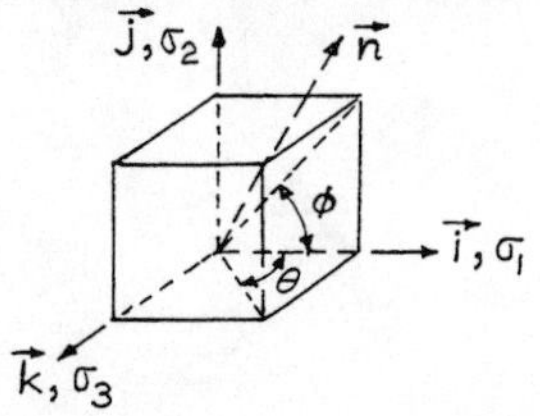

(CONT.)

1.54 CONT.

$\tan\theta = z/x$, $z = x\cdot\tan\theta$

$\tan\phi = y/x$, $y = x\cdot\tan\phi$

Also $\vec{n} = l\vec{i}+m\vec{j}+n\vec{k}$

$$\vec{n} = \frac{x\vec{i}+y\vec{j}+z\vec{k}}{x^2+y^2+z^2}$$

Equating $\vec{i},\vec{j},\vec{k}$ components of normal n:

$$l = x/[x^2+y^2+z^2]^{1/2}$$
$$= x/[x^2+x^2\cdot\tan^2\phi+x^2\cdot\tan^2\theta]^{1/2}$$
$$= 1/[1+\tan^2\phi+\tan^2\theta]^{1/2} = \sqrt{3/13}$$
$$m = y/[x^2+y^2+z^2]^{1/2}$$
$$= y/[x^2+x^2\cdot\tan^2\phi+x^2\cdot\tan^2\theta]^{1/2}$$
$$= \tan\phi/[1+\tan^2\phi+\tan^2\theta]^{1/2}$$
$$= l\tan\phi = \sqrt{1/13}$$
$$n = z/[x^2+y^2+z^2]^{1/2}$$
$$= \tan\theta/[1+\tan^2\phi+\tan^2\theta]^{1/2}$$
$$= l\tan\theta = \sqrt{9/13}$$

It is noted that $l^2+m^2+n^2=1$.

Applying Eq, (1.32), we have

σ =35(3/13)-14(1/13)-28(9/13)
=-12.39 MPa ◄

Equation (1.34), substituting the given data and the direction cosines determined above, gives

τ =26.2 MPa ◄

Surface tractions. Eqs. (1.41) give

$p_x = \sigma_1 l = 35(\sqrt{3/13})$
$= 16.81$ MPa

$p_y = \sigma_2 m = -14(\sqrt{1/13})$ ◄
$= -3.88$ MPa

$p_z = \sigma_3 n = -28(\sqrt{9/13})$
$= -23.30$ MPa

Check: $p^2 = p_x^2+p_y^2+p_z^2 = \sigma^2+\tau^2$
$= 840\ (\text{MPa})^2$

Observe that Approach I is more conveniently leads to results.

1.55

Approach I. Graphical Solution

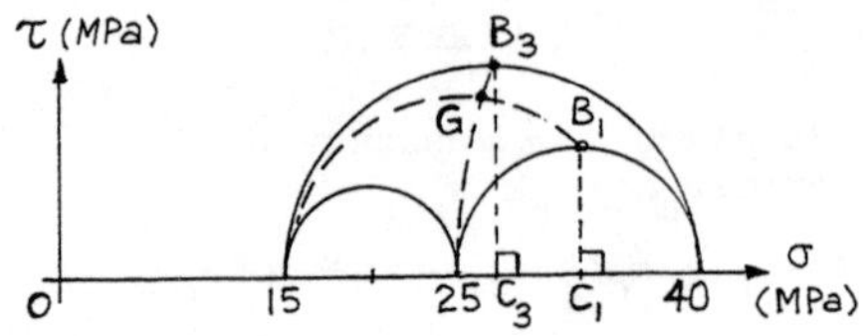

From Mohr's circle, we have

$\sigma_G = 26.67$ MPa $\tau_G = 10.27$ MPa ◄

Approach II. Analytical Solution

$\sigma_{oct} = \frac{1}{3}(40+25+15) = 26.667$ MPa ◄

$\tau_{oct} = \frac{1}{3}[(40-25)^2 + (25-15)^2 + (15-40)^2]^{1/2}$

$= 10.274$ MPa ◄

Then, referring to Solution of Problem 1.54:

$l = 1/[1+\tan^2 45° + \tan^2 45°]$

$= 1/\sqrt{3}$

$m = l \tan\phi = 1/\sqrt{3}$

$n = l \tan\theta = 1/\sqrt{3}$

Note that $l^2+m^2+n^2=1$.

Surface tractions. Eqs. (1.41) give

$p_x = \sigma_1 l = 40(1/\sqrt{3}) = 23.09$ MPa

$p_y = \sigma_2 m = 25(1/\sqrt{3}) = 14.43$ MPa ◄

$p_z = \sigma_3 n = 15(1/\sqrt{3}) = 8.66$ MPa

Check:

$p^2 = p_x^2 + p_y^2 + p_z^2 = \sigma^2 + \tau^2$

$= 816.7\ (\text{MPa})^2$

2.1

(a) Yes. ◄
Eqs. (2.9) are satisfied.
(b) No. ◄
Eqs. (2.9) are not satisfied.

2.2

Apply Eqs. (2.3):

$\varepsilon_x=2c \quad \varepsilon_y=-6cy \quad \gamma_{xy}=2c(x+y)$

(a) $u_{AB}=\int_1^3 \varepsilon_x dx=4c=0.4$ mm

$v_{AD}=\int_{1/2}^2 \varepsilon_y dy=-6c\frac{y^2}{2}\Big|_{1/2}^2=-11.25c$
$=-1.125$ mm

Thus,

$L_{A'B'}=2000.4$ mm ◄

$L_{A'D'}=1498.875$ mm

(b) $\gamma_{xy}=2c(1+\frac{1}{2})=300\mu$ ◄

(c) We have

$u_A=c(2\times1+\frac{1}{4})=2.25c$
$v_A=c(1^2-3\times\frac{1}{4})=0.25c$

and

$x_{A'}=1+2.25c=1000.225$ mm ◄
$y_{A'}=0.5+0.25c=500.025$ mm

2.3

Equations (2.3), for the given displacement field, yield $[\varepsilon_{ij}]$:

$$\begin{bmatrix} 2x & 0 & -y/2 \\ 0 & 2z & (2y-x)/2 \\ -y/2 & (2y-x)/2 & 2z \end{bmatrix} c$$

At point (0,2,1), we have $[\varepsilon_{ij}]$:

$$\begin{bmatrix} 0 & 0 & -100 \\ 0 & 200 & 200 \\ -100 & 200 & 200 \end{bmatrix} \mu$$ ◄

2.4

First two of Eqs. (2.3) give

$\varepsilon_x=2a_0xy^2+a_1y^2+2a_2xy$
$\varepsilon_y=b_0x^2+b_1x$

Equation (2.8):

$(4a_0+2a_1)+(2b_0)=2c_0x+c_1$

or

$2(2a_0-c_0)x+2(a_1+b_0)-c_1=0$

This is satisfied if $x \neq 0$:

$2a_0-c_0=0, \qquad c_0=2a_0$
$2(a_1+b_0)-c_1=0, \qquad c_1=2(a_1+b_0)$ ◄

2.5

Equation (2.8) yields

$$2a_1+12y^2+2b_1+12x^2=3c_1(x^2+y^2)+c_1c_2$$

Solving,

$c_1=4 \qquad c_2=\frac{1}{2}(a_1+b_1)$ ◄

2.6

(a) Equations (2.3) give

$\varepsilon_x=\partial u/\partial x=(0.175-0.075)/150$
$=667\mu$ ◄
$\varepsilon_y=\partial v/\partial y=[0.025-(-0.05)]/100$
$=750\mu$ ◄

and

$\gamma_{xy}=[(0-0.075)/100]+$
$[-0.125-(-0.05)]/150$
$=-1250\mu$ ◄

(b) Equation (2.13) is therefore

$$\varepsilon_{1,2}=\frac{667+750}{2}\pm\left[\left(\frac{667-750}{2}\right)^2+625^2\right]^{1/2}$$

or

$\varepsilon_1=1335\mu \quad \varepsilon_2=82\mu$ ◄

When

$$\theta_p=\frac{1}{2}\tan^{-1}\frac{-1250}{667-750}=43.1°$$

and stresses are substituted into Eq. (2.11a), we obtain $\varepsilon_{x'}=82\mu$. Thus,

$\theta_p''=43.1°$ ◄

2.7

Use Eq. (2.13) with the stresses obtained in Example 2.1:

$$\varepsilon_{1,2}=\frac{1250-2000}{2}\pm\left[\left(\frac{1250+2000}{2}\right)^2+750^2\right]^{1/2}$$

or

$\varepsilon_1=1415\mu \quad \varepsilon_2=-2165\mu$ ◄

Apply Eq. (2.12):

$$\theta_p=\frac{1}{2}\tan^{-1}\frac{1500}{1250+2000}=12.39°$$

Substituting this angle and the stresses, Eq. (2.11a) gives 1415μ. Thus,

$\theta_p'=12.39°$ ◄

2.8

We have, using Eqs. (2.3):

$$\varepsilon_x = \partial u/\partial x = 0.004/20 = 200\mu$$
$$\varepsilon_y = \partial v/\partial y = -0.003/12 = -250\mu$$

and

$$\gamma_{xy} = \frac{\partial v}{\partial x} + \frac{\partial u}{\partial y} = -1000 - 500 = -1500\mu$$

Then, Eq. (2.13) yields

$$\varepsilon_{1,2} = \frac{200-250}{2} \pm \left[\left(\frac{200+250}{2}\right)^2 + (-750)^2\right]^{1/2}$$

or

$$\varepsilon_1 = 758\mu \qquad \varepsilon_2 = -808\mu$$ ◄

Apply Eq. (2.12):

$$\theta_p = \frac{1}{2}\tan^{-1}\frac{-1500}{200+250} = -36.65°$$

For this angle, Eq. (2.11a) gives $\varepsilon_{x'} = -808\mu$. Therefore,

$$\theta_p'' = -36.65°$$ ◄

2.9

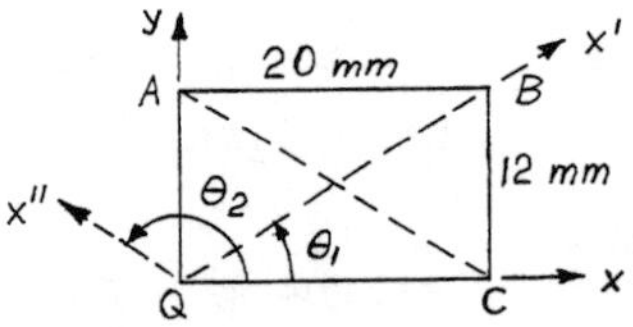

We have, from geometry:

AC=QB=23.32 mm

$\theta_1 = 30.96°$ $\theta_2 = 149.04°$

Equation (2.11a) leads to

$$\varepsilon_{x'} = \frac{300+500}{2} + \frac{300-500}{2}\cos 61.92° + 100\sin 61.92° = 441\mu$$

Thus,

$$\Delta_{QB} = 441\mu(23.32) = 0.01 \text{ mm}$$ ◄

Similarly,

$$\varepsilon_{x''} = 400 - 100\cos 298.08° + 100\sin 298.08° = 359\mu$$

and

$$\Delta_{AC} = 359\mu(23.32) = 0.008 \text{ mm}$$ ◄

2.10

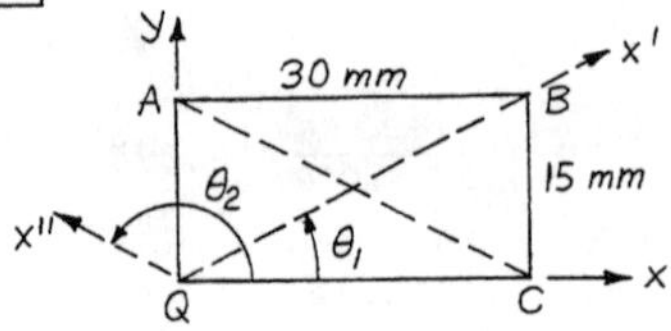

(CONT.)

2.10 CONT.

We find, from geometry:

AC=QB=33.54 mm

$\theta_1 = 26.56°$ $\theta_2 = 153.44°$

Equation (2.13) gives

$$\varepsilon_{x'} = \frac{400+200}{2} + \frac{400-200}{2}\cos 53.12° - 150\sin 53.12° = 240\mu$$

Hence,

$$\Delta_{QB} = 240\mu(33.54) = 0.008 \text{ mm}$$ ◄

In a like manner,

$$\varepsilon_{x''} = 300 + 100\cos 306.88° - 150\sin 306.88° = 120\mu$$

and

$$\Delta_{AC} = 120\mu(33.54) = 0.004 \text{ mm}$$ ◄

2.11

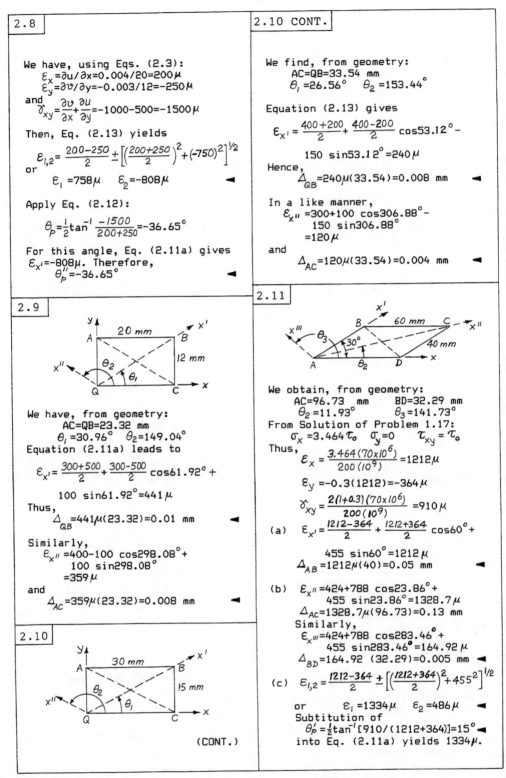

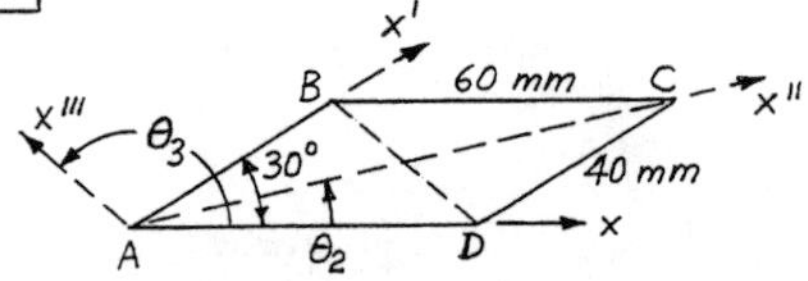

We obtain, from geometry:

AC=96.73 mm BD=32.29 mm

$\theta_2 = 11.93°$ $\theta_3 = 141.73°$

From Solution of Problem 1.17:

$\sigma_x = 3.464\tau_o$ $\sigma_y = 0$ $\tau_{xy} = \tau_o$

Thus,

$$\varepsilon_x = \frac{3.464(70\times10^6)}{200(10^9)} = 1212\mu$$

$$\varepsilon_y = -0.3(1212) = -364\mu$$

$$\gamma_{xy} = \frac{2(1+0.3)(70\times10^6)}{200(10^9)} = 910\mu$$

(a) $$\varepsilon_{x'} = \frac{1212-364}{2} + \frac{1212+364}{2}\cos 60° + 455\sin 60° = 1212\mu$$

$$\Delta_{AB} = 1212\mu(40) = 0.05 \text{ mm}$$ ◄

(b) $$\varepsilon_{x''} = 424 + 788\cos 23.86° + 455\sin 23.86° = 1328.7\mu$$

$$\Delta_{AC} = 1328.7\mu(96.73) = 0.13 \text{ mm}$$

Similarly,

$$\varepsilon_{x'''} = 424 + 788\cos 283.46° + 455\sin 283.46° = 164.92\mu$$

$$\Delta_{BD} = 164.92\,(32.29) = 0.005 \text{ mm}$$ ◄

(c) $$\varepsilon_{1,2} = \frac{1212-364}{2} \pm \left[\left(\frac{1212+364}{2}\right)^2 + 455^2\right]^{1/2}$$

or $\varepsilon_1 = 1334\mu$ $\varepsilon_2 = 486\mu$ ◄

Subtitution of

$$\theta_p' = \frac{1}{2}\tan^{-1}[910/(1212+364)] = 15°$$ ◄

into Eq. (2.11a) yields 1334μ.

2.12

(a) We have

$$\gamma_{max} = 400 - 200 = 200\,\mu$$ ◄

Maximum shearing strain occurs on a plane oriented at 45° from the plane of principal strains.

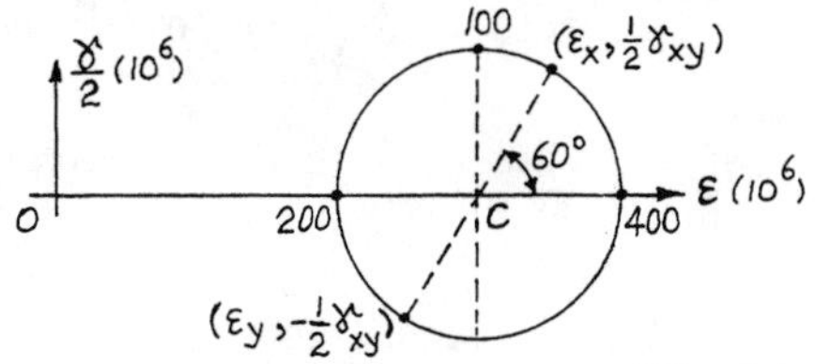

Fig. (a)

(b) From Mohr's circle (Fig. a):

$$\varepsilon_x = 300 + 100\cos 60° = 350\,\mu$$
$$\varepsilon_y = 300 - 100\cos 60° = 250\,\mu$$ ◄
$$\gamma_{xy} = -(400-200)\sin 60° = -173\,\mu$$

2.13

We have $u_B = 3$ mm and $v_B = 1.5$ mm.

(a) Take: $u = c_1 xy \qquad v = c_2 xy$

Hence,

$$3(10^{-3}) = c_1(3\times 2)$$

or

$$c_1 = 500(10^{-6})$$

$$1.5(10^{-3}) = c_2(3\times 2)$$

or

$$c_2 = 250(10^{-6})$$

Thus,

$$u = 500(10^{-6})xy$$
$$v = 250(10^{-6})xy$$ ◄

(b) Using Eqs. (2.3),

$$\varepsilon_x = 500\mu y \qquad \varepsilon_y = 250\mu x$$
$$\gamma_{xy} = 250\mu(2x+y)$$

which satisfy Eq. (2.8): the strain field is possible.

At point B, we thus have

$$(\varepsilon_x)_B = 1000\mu \qquad (\varepsilon_y)_B = 750\mu$$ ◄
$$(\gamma_{xy})_B = 2000\mu$$

(c) We obtain $\theta = \tan^{-1}(2/3) = 33.69°$. Equation (2.11a) is therefore

$$\varepsilon_{x'} = 875 + 125\cos 67.38° + 1000\sin 67.38°$$
$$= 184\mu$$ ◄

2.14

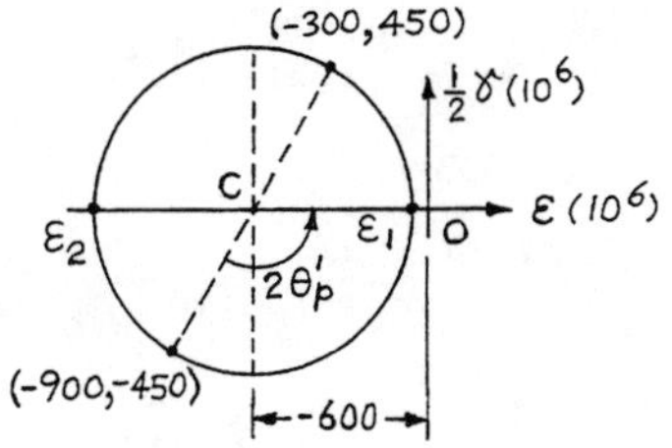

Fig. (a)

Refer to Mohr's circle (Fig. a):

$$\varepsilon_{1,2} = -600 \pm [(300)^2 + (450)^2]^{1/2}$$

or

$$\varepsilon_1 = -60\,\mu \qquad \varepsilon_2 = -1140\,\mu$$ ◄

and

$$2\theta_p'' = \tan^{-1}\frac{450(2)}{-(900-300)} = -56.31°$$

or

$$\theta_p' = 61.85°$$ ◄

2.15

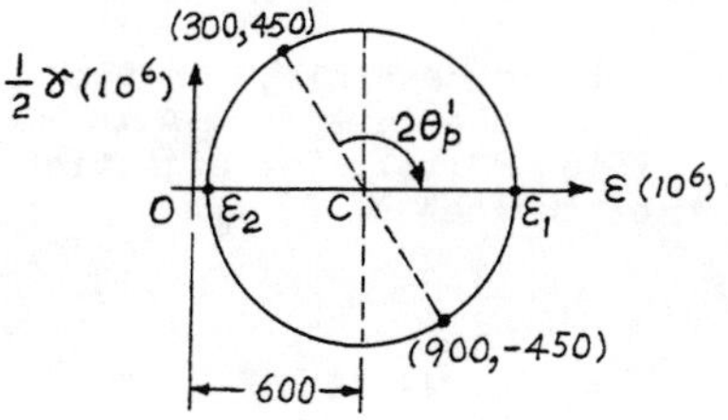

Fig. (a)

Referring to Mohr's circle shown in Fig. (a), we obtain

$$\varepsilon_{1,2} = 600 \pm [(300)^2 + (450)^2]^{1/2}$$

from which

$$\varepsilon_1 = 1140\mu \qquad \varepsilon_2 = 60\mu$$ ◄

and

$$2\theta_p'' = \tan^{-1}\frac{450(2)}{900-300} = 56.31°$$

or

$$\theta_p' = -61.85°$$ ◄

2.16

Referring to Fig. P2.16,

$u_A = -0.0005 = c_1(3\times1\times2)$, $c_1 = -83.3(10^{-6})$

$v_A = 0.0003 = c_2(6)$, $c_2 = 50(10^{-6})$

$w_A = -0.0006 = c_3(6)$, $c_3 = -100(10^{-6})$

Hence,

$u = -83.3(10^{-6})xyz$ $\quad v = 50(10^{-6})xyz$

$w = -100(10^{-6})xyz$

(a) Using Eqs. (2.3), we thus have

$$\varepsilon_x = -83.3\mu\, yz$$
$$\varepsilon_y = 50\mu\, xz$$
$$\varepsilon_z = -100\mu\, xy$$
$$\gamma_{xy} = (-83.3xz + 50yz)\mu$$
$$\gamma_{xz} = (-100yz - 83.4xy)\mu$$
$$\gamma_{yz} = (50xy - 100xz)\mu$$

The foregoing expressions satisy Eqs. (2.9): the strain field is possible.

Substitute x=3 m, y=1 m, and z=2 m into the above equations to obtain strains at A, $[\varepsilon_{ij}]$:

$$\begin{bmatrix} -167 & -200 & -225 \\ -200 & 300 & -225 \\ -225 & -225 & -300 \end{bmatrix}\mu \quad \blacktriangleleft$$

(b) Let, for example, x'-axis lie along the line from A to B (Fig. P2.16). The direction of cosines of AB are

$$l_1 = -\frac{3}{\sqrt{13}} \qquad m_1 = 0 \qquad n_1 = -\frac{2}{\sqrt{13}}$$

Thus, Eq. (2.15a):

$$\varepsilon_{x'} = \varepsilon_x l_1^2 + \varepsilon_z n_1^2 + \gamma_{xz} l_1 n_1$$

$$= -167(9/13) - 300(4/13) - 450(-3/\sqrt{13})(-2/\sqrt{13})$$
$$= -416\mu \quad \blacktriangleleft$$

(c) Let y'-axis be placed along AC (Fig. P2.16). The direction cosines of AC are

$$m_2 = -1 \qquad l_2 = 0 \qquad n_2 = 0$$

Equation (2.15b) is therefore

$$\gamma_{x'y'} = \gamma_{xy} l_1 m_2 + \gamma_{yz} n_1 m_2$$

$$= -400(-3/\sqrt{13})(-1) - 450(-2/\sqrt{13})(-1)$$
$$= -582\mu \quad \blacktriangleleft$$

Negative sign shows that the angle BAC has increased.

2.17

We now have (Fig. P2.16):

$u_A = 0.0006 = c_1(3\times1\times2)$, $c_1 = 100(10^{-6})$

$v_A = -0.0003 = c_2(6)$, $c_2 = -50(10^{-6})$

$w_A = -0.0004 = c_3(6)$, $c_3 = -66.7(10^{-6})$

Hence,

$u = 100(10^{-6})xyz$ $\quad v = -50(10^{-6})xyz$

$w = -66.7(10^{-6})xyz$

(a) Applying Eqs. (2.3), we obtain

$$\varepsilon_x = 100\mu\, yz$$
$$\varepsilon_y = -50\mu\, xz$$
$$\varepsilon_z = -66.7\mu\, xy$$
$$\gamma_{xy} = (100xz - 50yz)\mu$$
$$\gamma_{xz} = (-66.7yz + 100xy)\mu$$
$$\gamma_{yz} = (100xy - 66.7xz)\mu$$

These expressions satisfy Eqs. (2.9): the strain field is possible.

Introducing x=3 m, y=1, and z=2 m into the above equations we find strains at A, $[\varepsilon_{ij}]$:

$$\begin{bmatrix} 200 & 250 & 83.5 \\ 250 & -300 & -275 \\ 83.5 & -275 & -200 \end{bmatrix}\mu \quad \blacktriangleleft$$

(b) Let, for instance, x'-axis lie along the line from A to B (Fig. P2.16). The direction of cosines of AB are

$$l_1 = -\frac{3}{\sqrt{13}} \qquad m_1 = 0 \qquad n_1 = -\frac{2}{\sqrt{13}}$$

Therefore, Eq. (2.15a):

$$\varepsilon_{x'} = \varepsilon_x l_1^2 + \varepsilon_z n_1^2 + \gamma_{xz} l_1 n_1$$

$$= 200(9/13) - 200(4/13) + 167(-3/\sqrt{13})(-2/\sqrt{13})$$
$$= 154\mu \quad \blacktriangleleft$$

(c) Let y'-axis be placed along AC (Fig. P2.16). The direction cosines of AC are

$$m_2 = -1 \qquad l_2 = 0 \qquad n_2 = 0$$

Equation (2.15b) is thus

$$\gamma_{x'y'} = \gamma_{xy} l_1 m_2 + \gamma_{yz} n_1 m_2$$

$$= 500(-3/\sqrt{13})(-1) - 550(-2/\sqrt{13})(-1)$$
$$= 111\mu \quad \blacktriangleleft$$

Positive sign means that the angle BAC has decreased.

2.18

(a) Applying Eqs. (2.17),

$$J_1 = 200-100-400 = -300\,\mu$$

$$J_2 = (-2-8+4-9-4-25)(10^4) = -44(10^4)\ (\mu)^2$$

and

$$J_3 = \begin{vmatrix} 200 & 300 & 200 \\ 300 & -100 & 500 \\ 200 & 500 & -400 \end{vmatrix} = 58(10^6)\ (\mu)^3$$

(b) Table of direction cosines:

	x	y	z
x'	$\sqrt{3}/2$	1/2	0
y'	−1/2	$\sqrt{3}/2$	0
z'	0	0	1

Thus, using Eqs. (2.15a),

$$\varepsilon_{x'} = \varepsilon_x l_1^2 + \varepsilon_y m_1^2 + \gamma_{xy} l_1 m_1$$

$$= 200(\sqrt{3}/2)^2 - 100(1/2)^2 + 600(\sqrt{3}/2)(1/2) = 385\,\mu$$

(c) Use Table B.1 (with $\sigma \to \varepsilon$ and $\tau \to \gamma/2$):

$$\varepsilon_1 = 598\,\mu \qquad \varepsilon_2 = -126\,\mu \qquad \varepsilon_3 = -772\,\mu$$

(d) $\gamma_{max} = 598+772 = 1370\,\mu$

2.19

(a) Applying Eqs. (2.17):

$$J_1 = 400+0+600 = 1(10^3)\,\mu$$

$$J_2 = (0+24+0-1-4-0)(10^3)\,\mu = 19(10^4)\ (\mu)^2$$

and

$$J_3 = \begin{vmatrix} 400 & 100 & 0 \\ 100 & 0 & -200 \\ 0 & -200 & 600 \end{vmatrix} = -22(10^6)\ (\mu)^3$$

(b) Using Eq. (2.15a),

$$\varepsilon_{x'} = 400(\sqrt{3}/2)^2 + 200(\sqrt{3}/2)(1/2) = 387\,\mu$$

(c) Use Table B.1 (with $\sigma \to \varepsilon$ and $\tau \to \gamma/2$):

$$\varepsilon_1 = 664\,\mu \qquad \varepsilon_2 = 416\,\mu \qquad \varepsilon_3 = -80\,\mu$$

(d) $\gamma_{max} = 664+80 = 744\,\mu$

2.20

Use Table B.1 (with $\sigma \to \varepsilon$ and $\tau \to \gamma/2$):

$$\varepsilon_1 = 1807\,\mu \qquad \varepsilon_2 = -228\,\mu \qquad \varepsilon_3 = -679\,\mu$$

and

$$l_1 = 0.6184 \qquad m_1 = 0.5333 \qquad n_1 = 0.5772$$

2.21

Nominal strain

$$\varepsilon_0 = 0.025/75 = 333\,\mu$$

Nominal stress

$$\sigma_0 = \frac{9(10^3)}{\pi(0.012)^2/4} = 79.577 \text{ MPa}$$

Modulus of elasticity

$$E = \frac{79.577(10^6)}{333(10^{-6})} = 238.97 \text{ GPa}$$

True strain

$$\varepsilon = \ln(1+0.000333) = 333\,\mu$$

True stress

$$\sigma = 79.577(1+0.000333) = 79.603 \text{ MPa}$$

2.22

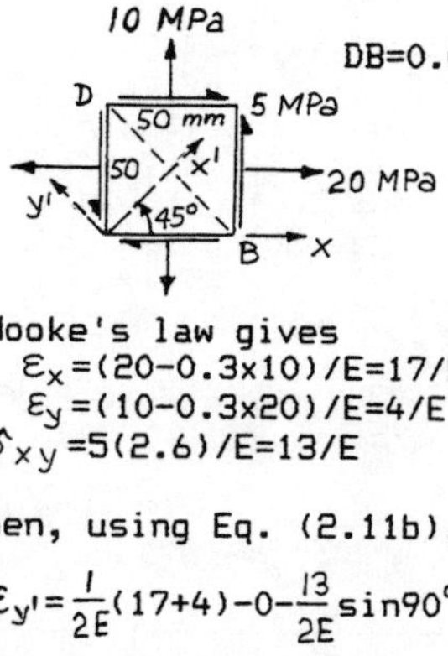

$DB = 0.005\sqrt{2}$ m

(a) Hooke's law gives

$$\varepsilon_x = (20-0.3\times10)/E = 17/E$$
$$\varepsilon_y = (10-0.3\times20)/E = 4/E$$
$$\gamma_{xy} = 5(2.6)/E = 13/E$$

Then, using Eq. (2.11b),

$$\varepsilon_{y'} = \frac{1}{2E}(17+4) - 0 - \frac{13}{2E}\sin 90° = \frac{4}{E}$$

Thus,

$$\Delta_{BD} = (\varepsilon_{y'})(BD) = 0.283/E \text{ m}$$

(b) Apply Eqs. (2.11a,b):

$$\sigma_{x'} = \frac{30}{2} + \frac{10}{2}\cos 90° + 5\sin 90° = 20 \text{ MPa}$$

$$\sigma_{y'} = 15 - 5\cos 90° - 5\sin 90° = 10 \text{ MPa}$$

As before, Hooke's law yields

$$\varepsilon_{y'} = (10-0.3\times20)/E = 4/E$$

and

$$\Delta_{BD} = (\varepsilon_{y'})(BD) = 0.283/E \text{ m}$$

2.23

Using Eqs. (2.36):

$$\varepsilon_{1,2}=\frac{1}{2}\{(-100+100)\pm[(-200)^2+100^2]^{1/2}\}$$
$$=\frac{1}{2}(0\pm 224)$$

or

$$\varepsilon_1=112\mu \qquad \varepsilon_2=-112\mu$$ ◄

$$\sigma_{1,2}=\frac{200(10^3)}{2}[0\pm(224/1.3)]$$

or

$$\sigma_1=17.2\text{ MPa} \qquad \sigma_2=-17.2\text{ MPa}$$ ◄

and

$$\theta_p=\frac{1}{2}\tan^{-1}\frac{100}{-200}=-13.28^\circ$$

Equations (2.34),

$$\varepsilon_x=\varepsilon_a \qquad \varepsilon_y=\varepsilon_c$$
$$\gamma_{xy}=2\varepsilon_b-\varepsilon_a-\varepsilon_c$$

Equation (2.11a) gives

$$\varepsilon_{x'}=0-100\cos(-26.56^\circ)+50\sin(-26.56^\circ)=-112\mu$$

Thus,

$$\theta_p''=-13.28^\circ$$ ◄

2.24

We have

$$\varepsilon_x+\varepsilon_y=\varepsilon_a+\varepsilon_c=1200\mu$$

and the first two of Eqs. (2.34):

$$1000=\varepsilon_x\cos^2(-15^\circ)+\varepsilon_y\sin^2(-15^\circ)+\gamma_{xy}\sin(-15^\circ)\cos(-15^\circ)$$
$$-250=\varepsilon_x\cos^2 30^\circ+\varepsilon_y\sin^2 30^\circ+\gamma_{xy}\sin 30^\circ\cos 30^\circ$$

These may be written

$$\varepsilon_y=1200-\varepsilon_x$$
$$1000=0.933\varepsilon_x+0.067\varepsilon_y-0.25\gamma_{xy}$$
$$-250=0.75\varepsilon_x+0.25\varepsilon_y+0.433\gamma_{xy}$$

Solving,

$$\varepsilon_x=522\mu \qquad \varepsilon_y=678\mu$$
$$\gamma_{xy}=-1873\mu$$ ◄

2.25

(a) We have

$$\varepsilon_c=\varepsilon_y=-50\mu$$ ◄

First two of Eqs. (2.34):

$$400=\varepsilon_x(3/4)-50(1/4)+\gamma_{xy}(1/2)(\sqrt{3}/2)$$
$$300=\varepsilon_x(3/4)-50(1/4)+\gamma_{xy}(-1/2)(\sqrt{3}/2)$$

Solving,

$$\varepsilon_x=483\mu \qquad \gamma_{xy}=116\mu$$ ◄

(CONT.)

2.25 CONT.

Applying Eq. (2.13),

$$\varepsilon_{1,2}=\frac{483-50}{2}\pm\left[\left(\frac{483+50}{2}\right)^2+58^2\right]^{1/2}$$

or

$$\varepsilon_1=489\mu \qquad \varepsilon_2=-56\mu$$

Thus,

$$\gamma_{max}=\varepsilon_1-\varepsilon_2=545\mu$$ ◄

(b) Using Eq. (3.11b),

$$\varepsilon_z=-\frac{\nu}{1-\nu}(483-50)=-217\mu=\varepsilon_3$$

Hence,

$$(\gamma_{max})_t=\varepsilon_1-\varepsilon_3=706\mu$$ ◄

2.26

From Eqs. (2.29), (2.31), and (2.32), we have

$$\nu=\frac{200(10^9)}{2(80\times10^9)}-1=0.25$$
$$=\frac{0.25\times200(10^9)}{1.25\times0.5}=80(10^9)$$

and

$$e=200+300=500\mu$$

Then, Eqs. (2.30) lead to the following stress components, $[\tau_{ij}]$:

$$\begin{bmatrix}72 & 16 & 0\\ 16 & 88 & 64\\ 0 & 64 & 40\end{bmatrix}\text{ MPa}$$ ◄

2.27

Equation (2.29) yields

$$\nu=\frac{200(10^9)}{2(80\times10^9)}-1=0.25$$

Then, introducing the given data into the generalized Hooke's law, Eqs. (2.28), we calculate the following strain components, $[\varepsilon_{ij}]$:

$$\begin{bmatrix}81.25 & -25 & 31.25\\ -25 & -43.75 & 62.5\\ 31.25 & 62.5 & 50\end{bmatrix}\mu$$ ◄

2.28

Using Eq. (2.29),
$G=79(10^9)/2(1+0.3)=26.92$ GPa
For x=1 m, y=2 m, and z=4 m we obtain $[\tau_{ij}]$:

$$\begin{bmatrix} 34 & 48 & 8 \\ 48 & 5 & 1 \\ 8 & 1 & 5 \end{bmatrix} \text{ MPa}$$

Then, Eqs. (2.28) yields $[\varepsilon_{ij}]$:

$$\begin{bmatrix} 443 & 892 & 149 \\ 892 & -96 & 19 \\ 149 & 19 & -96 \end{bmatrix} \mu$$ ◄

2.29

Substituting x=3/4 m, y=1/4 m, and z=1/2 m into Eqs. (d) of Example 1.2, we have $[\tau_{ij}]$:

$$\begin{bmatrix} -0.359 & 2.625 & 0.234 \\ 0.234 & 0.875 & 0 \\ 0.234 & 0 & 0.125 \end{bmatrix} \text{ MPa}$$

Equation (2.29) gives,
$G=200(10^9)/2(1+0.25)=80$ GPa

Applying Hooke's law, we compute the strain components $[\varepsilon_{ij}]$:

$$\begin{bmatrix} -3 & 33 & 3 \\ 33 & 5 & 0 \\ 3 & 0 & 0 \end{bmatrix} \mu$$ ◄

2.30

(a) Using generalized Hooke's law,

$$\varepsilon_x = \frac{10^6}{200(10^9)}[-60-0.3(-50-40)]$$
$$=-165\mu$$

$$\varepsilon_y = \frac{1}{200(10^3)}[-50-0.3(-60-40)]$$
$$=-100\mu$$

$$\varepsilon_z = \frac{1}{200(10^3)}[-40-0.3(-60-50)]$$
$$=-35\mu$$

Thus,

$\Delta a=a\varepsilon_x=-0.04125$ mm
$\Delta b=b\varepsilon_y=-0.02$ mm ◄
$\Delta c=c\varepsilon_z=-0.00525$ mm

(b) $e=\varepsilon_x+\varepsilon_y+\varepsilon_z=-300\mu$
and
$\Delta V=e(abc)=-2250$ mm^3 ◄

2.31

Applying generalized Hooke's law,

$$\varepsilon_x = \frac{10^6}{70(10^9)}[70-\tfrac{1}{3}(-30-15)]=1214\mu$$

$$\varepsilon_y = \frac{1}{70(10^3)}[-30-\tfrac{1}{3}(70-15)]=-691\mu$$

$$\varepsilon_z = \frac{1}{70(10^3)}[-15-\tfrac{1}{3}(70-30)]=-405\mu$$

Thus,

$\Delta a=a\varepsilon_x=0.1821$ mm
$\Delta b=b\varepsilon_y=-0.0691$ mm ◄
$\Delta c=c\varepsilon_z=-0.0304$ mm

(b) $e=\varepsilon_x+\varepsilon_y+\varepsilon_z=118\mu$
and
$\Delta V=e(abc)$
$=132.75$ mm^3 ◄

2.32

We have
$G=200(10^9)/2(1+0.3)=76.96$ GPa

$$\lambda = \frac{0.3\times 200(10^9)}{1.3(0.4)}=111.11(10^9)$$

Let $\varepsilon_1=5c$ $\varepsilon_2=4c$ $\varepsilon_3=3c$
We then have e=12c.

Using the first of Eqs. (2.30),
$\sigma_1=2G\varepsilon_1+\lambda e$
or
$140(10^6)=2(76.96\times 10^9)(5c)+111.11(10^9)(12c)$
This yields
$c=66.587(10^{-6})$
Hence,
$\varepsilon_1=332.935\mu$ $\varepsilon_2=266.348\mu$
$\varepsilon_3=199.761\mu$

Now applying Eqs. (2.30), we calculate the principal stresses as follows:

$\sigma_1=51.165+88.782=139.947$ MPa

$\sigma_2=40.975+88.782=129.757$ MPa

$\sigma_3=30.731+88.782=119.513$ MPa

Therefore,

$\sigma_3:\sigma_2:\sigma_1=119.513:129.757:139.947$

or

$\sigma_3:\sigma_2:\sigma_1=1:1.086:1.171$ ◄

2.33

Equations (2.28) become

$\varepsilon_x=(cy^2+\nu cx^2)/E$

$\varepsilon_y=(-cx^2+\nu cy^2)/E$

Integrating,

$$u(x,y)=\int\varepsilon_x dx=\frac{c}{3E}(3y^2x+\nu x^3)+g_1(y) \qquad (1)$$

$$v(x,y)=\int\varepsilon_y dy=\frac{c}{3E}(3x^2y+\nu y^2)+g_2(x) \qquad (2)$$

Given:

$$\tau_{xy}=0 \qquad \gamma_{xy}=\frac{\partial u}{\partial y}+\frac{\partial v}{\partial x}=0 \qquad (3)$$

From Eq. (1),

$$\frac{\partial u}{\partial y}=\frac{dg_1}{dy}+2\frac{c}{E}xy \qquad (a)$$

Equation (2) leads to

$$\frac{\partial v}{\partial x}=\frac{dg_2}{dx}-2\frac{c}{E}xy \qquad (b)$$

Substituting Eqs. (a) and (b) into Eq. (3):

$$\frac{dg_1}{dy}=2\frac{c}{E}xy+a_1 \qquad \frac{dg_2}{dx}=-2\frac{c}{E}xy+a_1$$

from which, after integration,

$$g_1(y)=\frac{c}{E}xy^2+a_1y+a_2$$
$$g_2(x)=-\frac{c}{E}yx^2-a_1x-a_2 \qquad (4)$$

Constants a_1 and a_2 are obtained upon satisfying the prescribed boundary conditions.

Substitution of Eqs. (4) into Eqs. (1) and (2) yields the solution for displacement field.

2.34

Equations (2.33) and (2.29) give

$$K=\frac{E}{3(1-2\nu)}=\frac{2G(1+\nu)}{3(1-2\nu)}$$

Also,

$$K=\frac{E}{3(1-2\nu)}=\frac{2E(1+\nu)}{3(1+\nu)(1-2\nu)}$$

$$=\frac{E(3\nu+1-2\nu)}{3(1+\nu)(1-2\nu)}=\frac{\nu E}{(1+\nu)(1-2\nu)}+\frac{E}{3(1+\nu)}$$

$$=\lambda+(2G/3) \quad ◄$$

(CONT.)

2.34 CONT.

Equations (2.33) and (2.29) yield

$$G=\frac{3K(1-2\nu)}{2(1+\nu)}=\frac{E/(1-2\nu)}{(3-1+2\nu)/(1-2\nu)}$$

$$=\frac{E^2/(1-2\nu)}{E\{[3/(1-2\nu)]-1\}}$$

$$=\frac{\frac{3E^2}{3E(1-2\nu)}}{\frac{9E}{3}\frac{1}{1-2\nu}-E}=\frac{3KE}{9K-E} \quad ◄$$

Equations (2.29) and (2.33) give,

$$E=2G(1+\nu), \qquad E=3K(1-2E)$$

and

$$\frac{G(3+2G)}{\lambda+G}=\frac{\frac{E}{2(1+\nu)}[3\lambda+\frac{E}{1+\nu}]}{\lambda+\frac{E}{2(1+\nu)}}$$

The above expression, after substituting λ from Eq. (2.32) and simplifying, reduce to E.

From formula (2.33),

$$(1-2\nu)=E/3K$$

or

$$\nu=\frac{1}{2}-\frac{E}{6K}=\frac{3K-E}{6K}$$

We also have

$$G=E/2(1+\nu) \qquad (1+\nu)=E/2G$$

or

$$\nu=\frac{E}{2G}-1 \quad ◄$$

This expression is written as

$$\nu=\frac{3K(1-2\nu)}{2G}-1=\frac{3K}{2G}-\frac{6K\nu}{2G}-1$$

from which

$$\nu+\frac{6K\nu}{2G}=\nu\left(1+\frac{6\nu}{2G}\right)=\frac{3K}{2G}-1$$

or

$$\nu=\frac{3K-2G}{2(3K+G)}$$

Finally, we write

$$\lambda=\frac{2\nu E}{2(1+\nu)(1-2\nu)}=\frac{2\nu G}{1-2\nu}$$

or

$$(1-2\nu)\lambda-2\nu G=0$$

This yields,

$$\nu=\frac{\lambda}{2(\lambda+G)} \quad ◄$$

2.35

Equations (2.3) yield,

$$\varepsilon_x = \gamma(a-x)/E$$
$$\varepsilon_y = -\nu\gamma(a-x)/E$$

and

$$\gamma_{xy} = -\nu\gamma y/E + \nu\gamma y/E = 0$$

The stresses are therefore,

$$\sigma_x = \frac{E}{1-\nu^2}(\varepsilon_x + \nu\varepsilon_y) = \gamma(a-x)$$

$$\sigma_y = \frac{E}{1-\nu^2}(\varepsilon_y + \nu\varepsilon_x) = 0 \quad ◄$$

$$\tau_{xy} = G\gamma_{xy} = 0$$

At x=0 and x=a, we have

$$\sigma_x = \gamma a \qquad \sigma_x = 0 \quad ◄$$

Applying Eqs. (1.41) at x=a, we obtain

$$\begin{Bmatrix} p_x \\ p_y \\ p_z \end{Bmatrix} = \begin{bmatrix} 0 & 0 & 0 \\ 0 & 0 & 0 \\ 0 & 0 & 0 \end{bmatrix} \begin{Bmatrix} 1 \\ 0 \\ 0 \end{Bmatrix} = 0$$

That is, boundary conditions at x=a are satisfied. ◄

At $y=\pm b$, stresses are

$$\sigma_x = \gamma(a-x) \qquad \sigma_y = \tau_{xy} = 0$$

and

$$\begin{Bmatrix} p_x \\ p_y \\ p_z \end{Bmatrix} = \begin{bmatrix} (x-a)\gamma & 0 & 0 \\ 0 & 0 & 0 \\ 0 & 0 & 0 \end{bmatrix} \begin{Bmatrix} 0 \\ \pm 1 \\ 0 \end{Bmatrix} = 0$$

We see that boundary conditions at $y=\pm b$ are also satisfied. ◄

2.36

Equations (2.3) give,

$$\varepsilon_x = -\nu\gamma z/E, \quad \varepsilon_y = -\nu\gamma z/E$$
$$\varepsilon_z = \gamma z/E, \quad \gamma_{xy} = \gamma_{yz} = \gamma_{xz} = 0 \qquad \text{(a)}$$

Equations (2.9) are satisfied by the above strains.

Equations (2.30) and (a) yield,

$$\sigma_x = 2G\varepsilon_x + \lambda(\varepsilon_x + \varepsilon_y + \varepsilon_z) = 0$$

and

$$\sigma_y = 0 \qquad \sigma_z = \gamma z$$
$$\tau_{xy} = \tau_{yz} = \tau_{xz} = 0$$

Note that

$$F_x = F_y = 0 \qquad F_z = -\gamma$$

Thus, the first of Eqs. (1.11) are satisfied, and the third leads to

$$0+0+(\partial\sigma_z/\partial z)+F_z = 0 \quad \text{or} \quad \gamma-\gamma=0$$

(CONT.)

2.36 CONT.

At ends, since x' and z are parallel, $n=\cos(x',z)=\pm 1$. Thus, boundary conditions (1.41) at z=L:

$$p_x = 0+0+0 = 0$$
$$p_y = 0+0+0 = 0$$
$$p_z = 0+0+\gamma L(1) = \gamma L$$

as required. Similarly, at z=0:

$$p_x = 0 \qquad p_y = 0 \qquad p_z = \gamma(0)(-1) = 0$$

Therefore, all equations of elasticity are satisfied. ◄

2.37

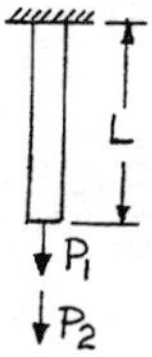

Using Eq. (2.49), with $P=P_1+P_2$:

$$U = \frac{(P_1+P_2)^2 L}{2AE} \quad ◄$$

From Example 2.8, with $P_1=P_2=P$:

$$U_v = \frac{\sigma^2}{12E}(AL) = \frac{P^2L}{3AE}$$

$$U_d = \frac{5\sigma^2}{12E}(AL) = \frac{5P^2L}{3AE} \quad ◄$$

Hence,

$$U = U_v + U_d = 2\frac{P^2L}{AE} \quad ◄$$

2.38

We have

$$U_1 = P^2L/2AE$$

$$U_2 = \frac{P^2(L/4)}{2AE} + \frac{P^2(3L/4)}{2E(2A)} = \frac{5}{8}U_1$$

and

$$U_3 = \frac{P^2(L/8)}{2AE} + \frac{P^2(7L/8)}{2E(3A)} = \frac{5}{12}U_1$$

Comparison of these results show that strain energy decreases as the volume of the bar is increased, although all three bars have the same maximum stress. ◄

2.39

Stress field is described by

$$\sigma_x=\sigma_y=\sigma_z=-p, \quad \tau_{xy}=\tau_{xz}=\tau_{yz}=0$$

(a) Equation (2.31) reduces to

$$e=-3(1-2\nu)p/E$$
$$-0.005=-3[1-(2/3)]p/110(10^9)$$

or

$$p=550 \text{ MPa}$$ ◄

(b) Equation (2.42) becomes

$$U_o=\frac{1}{2E}[p^2+p^2+p^2]-\frac{\nu}{E}(p^2+p^2+p^2)$$
$$=3p^2(1-2\nu)/2E$$
$$=\frac{3(550\times10^6)^2}{2(110\times10^9)}(1-\frac{2}{3})=1.375 \text{ MPa}$$

Thus,

$$U=U_oV_o=1.375(10^6)(\frac{4\pi}{3}\times0.15^3)$$
$$=19.439 \text{ kN}\cdot\text{m}$$ ◄

2.40

Substituting the given data into Eq. (2.42), we have

$$U_o=\frac{10^3}{2(200)}(60^2+50^2+40^2)-$$
$$\frac{0.3(10^3)}{200}[3000+2400+2000]$$
$$=8.15 \text{ kPa}$$

Thus,

$$U=U_o(abc)$$
$$=8.15(10^3)(0.25\times0.2\times0.15)$$
$$=61.125 \text{ N}\cdot\text{m}$$ ◄

2.41

Equation (2.49) leads to

$$U_n=\frac{P^2(3L/4)}{2AE}+\frac{P^2(L/4)}{2(n^2A)E}=\frac{P^2L}{8AE}(3+\frac{1}{n^2})$$
$$=\frac{1+3n^2}{4n^2}\frac{P^2L}{2AE}$$ ◄

We have,

for n=1: $U_1=P^2L/2AE$

Thus,

for n=1/2: $U_{1/2}=7U_1/4$

for n=2: $U_2=13U_1/16$

Hence,

$U_{1/2}>U_1$ and $U_2<U_1$ ◄

2.42

(a)
$$U=\frac{1}{2E}\int dx\int\sigma^2 dA=\frac{1}{2E}\int dx\int(\frac{P}{A}+\frac{My}{I})^2 dA$$

Since $\int ydA=0$, this becomes:

$$U=\int\frac{P^2dx}{2AE}+\int\frac{M^2dx}{2EI}=U_a+U_b$$ ◄

Here

$$U_b=\frac{1}{2EI}[\int_0^a M_{AD}^2dx+\int_0^b M_{DB}^2dx]$$
$$=\frac{1}{2EI}\left[\int_0^a\left(-\frac{M_ox}{L}\right)^2dx+\int_0^b\left(\frac{M_ox'}{L}\right)^2dx'\right]$$
$$=\frac{M_o^2}{6EIL^2}(a^3+b^3)$$ ◄

Thus,

$$U=\frac{P^2L}{2AE}+\frac{M_o^2}{6EIL^2}(a^3+b^3)$$

(b) Substituting the given data:

$$U=\frac{(8\times10^3)^2(1.2)}{2(7.5\times10^{-3})(70\times10^9)}+$$
$$\frac{(2\times10^3)^2(0.027+0.729)}{6(70\times10^9)(0.075\times0.1^3/12)(1.2)^2}$$
$$=0.0731+0.8=0.8731 \text{ N}\cdot\text{m}$$ ◄

2.43

We have for segment AB, $T_{AB}=3T$ and for segment BC, $T_{BC}=T$

(a)
$$U_{AB}=\left(\frac{T^2L}{2JG}\right)_{AB}=\frac{9T^2(1.2a)}{\pi(1.4d)^4G/16}$$
$$=44.977\frac{T^2a}{\pi d^4G}$$

and

$$U_{BC}=\left(\frac{T^2L}{2JG}\right)_{BC}=\frac{16T^2a}{\pi d^4G}$$

The total energy is thus

$$U=60.981\frac{T^2a}{\pi d^4G}$$ ◄

(b) Substituting the given data into the above equation,

$$U=60.981\frac{(1.4\times10^3)^2(0.5)}{\pi(0.02)^4(42\times10^9)}$$
$$=2.831 \text{ kN}\cdot\text{m}$$ ◄

2.44

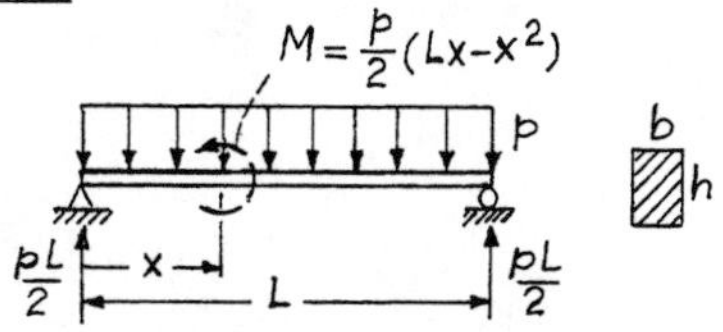

The maximum bending moment occurs at the midspan:

$$\sigma_{max}=\frac{M_{max}c}{I}=\frac{pL^2/8\ (h/2)}{bh^3/12}=\frac{3pL^2}{4bh^2}$$

Maximum strain energy density,

$$U_{o,max}=\sigma_{max}^2/2E=9p^2L^4/32Eb^2h^4$$

Using Eq. (2.53), we obtain

$$U=\frac{1}{2EI}\int_0^L(\frac{p}{2})^2(Lx-x^2)^2dx$$

$$=\frac{p^2L^5}{240EI}=\frac{p^2L^5}{20E\ bh^3}$$

It is required to find c:

$$U_{o,max}=cU/V$$

or

$$c=U_{o,max}\frac{V}{U}$$

Thus,

$$c=\frac{9p^2L^4}{32Eb^2h^4}\ \frac{bhL}{p^2L^5/20Ebh^3}=45/8$$

and

$$U_{o,max}=45U/8V$$ ◄

2.45

Equation (2.56) yields

$$\tau_{oct}=\frac{1}{3}[(-19-4.6)^2+(4.6+8.3)^2+(8.3+19)^2+6(4.7^2+6.45^2+1.8^2]^{1/2}=14.42\text{ MPa}$$

Applying Eqs. (2.55) and (2.54), respectively,

$$U_{od}=\frac{3}{4(76.9\times10^9)}(14.42)^2$$

$$=2027.99\text{ Pa}$$ ◄

and

$$U_{ov}=\frac{(-19+4.6-8.3)^2(10^{12})}{18(166.67\times10^9)}$$

$$=171.76\text{ Pa}$$ ◄

It follows that

$$U_{od}/U_{ov}=11.81\approx12$$ ◄

2.46

Dilatational stress is

$$\sigma_m=(200-50+40)/3=63.33\text{ MPa}$$

Distortional stresses are then

$$\sigma_x-\sigma_m=200-63.33=136.6\text{ MPa}$$
$$\sigma_y-\sigma_m=-50-63.33=-113.3\text{ MPa}$$
$$\sigma_z-\sigma_m=40-63.33=-23.33\text{ MPa}$$

Dilatational and deviator stress matrices are, respectively,

$$\begin{bmatrix}63.33 & 0 & 0\\ 0 & 63.33 & 0\\ 0 & 0 & 63.33\end{bmatrix}\text{ MPa}$$ ◄

and

$$\begin{bmatrix}136.6 & 20 & 10\\ 20 & -113.3 & 0\\ 10 & 0 & -23.33\end{bmatrix}\text{ MPa}$$ ◄

Next, substitute the above deviator stresses into Eq. (1.28) Then, solve the resulting equation for principal deviator stresses, using Table B.1:

$$\sigma_{1d}=138.8\text{ MPa}$$
$$\sigma_{2d}=-23.9\text{ MPa}$$
$$\sigma_{3d}=-114.9\text{ MPa}$$ ◄

2.47

Only existing stresses are:

$$\sigma_x=\sigma \qquad \tau_{xy}=\tau$$

Here,

$$\tau=\frac{2T}{\pi r^3}=\frac{2(20\times10^3)}{\pi(0.06)^3}=58.95\text{ MPa}$$

$$\sigma=\frac{4M}{\pi r^3}=\frac{4(15\times10^3)}{\pi(0.06)^3}=88.42\text{ MPa}$$

We have

$$G=2E/5 \qquad K=2E/3$$

Equations (2.54) and (2.55):

$$U_{ov}=\frac{\sigma^2}{18K}=\frac{\sigma^2}{12E}=\frac{(88.42)^2(10^3)}{12(200)}$$

$$=3.258\text{ kPa}$$ ◄

$$U_{od}=\frac{5}{12E}(\sigma^2+3\tau^2)$$

$$=\frac{5(10^3)}{12(200)}(7818.1+10{,}425.3)$$

$$=38.007\text{ kPa}$$ ◄

Thus,

$$U=U_{ov}+U_{od}=41.265\text{ kPa}$$ ◄

2.48

Principal stresses are

$$\sigma_{1,2} = \frac{\sigma}{2} \pm \left[\frac{\sigma^2}{4}+\tau^2\right]^{1/2}, \quad \sigma_3=0$$

and

$$\tau_{oct}=\frac{1}{3}(2\sigma^2+6\tau^2)^{1/2}$$

Equations (2.54) and (2.55) are therefore

$$U_{ov}=\frac{1}{18K}(\sigma_1^2+\sigma_2^2)=\frac{\sigma^2}{18K}=\frac{1-2\nu}{6E}\sigma^2 \quad \blacktriangleleft$$

$$U_{od}=\frac{3}{4G}\tau_{oct}^2=\frac{1+\nu}{3E}(\sigma^2+3\tau^2)$$

2.49

Only existing stress components are

$$\sigma_x = \sigma \qquad \tau_{xy} = \tau$$

where,

$$\sigma = P/\pi r^2 \qquad \tau = 2T/\pi r^3$$

The area properties are

$$A=\pi r^2 \qquad J=\pi r^4/2$$

Thus, Eqs. (2.54) and (2.55) become

$$U_{ov}=\frac{\sigma^2}{18K}=\frac{1-2\nu}{6E}\sigma^2=\frac{\sigma^2}{12E}$$

$$U_{od}=\frac{3}{4G}\tau_{oct}^2=\frac{5\sigma^2}{12E}+\frac{5\tau^2}{4E}$$

The components of strain energy are

$$U_v=\int U_{ov}dV$$

$$=\frac{1}{12E}\int\frac{P^2}{A}dx=\frac{P^2L}{12\pi r^2E} \quad \blacktriangleleft$$

$$U_d=\int U_{od}dV$$

$$=\frac{5}{12E}\int\frac{P^2}{E}dx+\frac{5}{4E}\int\frac{T^2}{J}dx$$

$$=\frac{5P^2L}{12\pi r^2E}+\frac{5T^2L}{2\pi r^4E} \quad \blacktriangleleft$$

Total strain energy is therefore

$$U=\frac{L}{2\pi r^2E}\left(P^2+5\frac{T^2}{r^2}\right) \quad \blacktriangleleft$$

3.1

(a) We obtain

$$\partial^4\Phi/\partial x^4=-12pxy \qquad \partial^4\Phi/\partial y^4=0$$
$$\partial^4\Phi/\partial x^2\partial y^2=6pxy$$

Thus,

$$\nabla^4\Phi=-12pxy+2(6pxy)=0$$

and the given stress field represents a possible solution. ◄

(b)

$$\partial^2\Phi/\partial x^2=pxy^3-2px^3y$$

Integrating twice

$$\Phi=\frac{px^3y^3}{6}-\frac{px^5}{10}y+f_1(y)x+f_2(y)$$

The above is substituted into $\nabla^4\Phi=0$ to obtain

$$\frac{d^4f_1(y)}{dy^4}x+\frac{d^4f_2(y)}{dy^4}=0$$

This is possible only if

$$\frac{d^4f_1(y)}{dy^4}=0 \qquad \frac{d^4f_2(y)}{dy^4}=0$$

We find then

$$f_1=c_4y^3+c_5y^2+c_6y+c_7$$
$$f_2=c_8y^3+c_9y^2+c_{10}y+c_{11}$$

Therefore,

$$\Phi=\frac{px^3y^3}{6}-\frac{px^5y}{10}+(c_4y^3+c_5y^2+c_6y+c_7)x+c_8y^3+c_9y^2+c_{10}y+c_{11}$$ ◄

(c) Edge y=0:

$$V_x=\int_{-a}^{a}\tau_{xy}t\,dx=\int_{-a}^{a}\left(\frac{px^4}{2}+c_4\right)t\,dx$$
$$=\frac{pa^5}{5}+2c_3at$$ ◄

$$P_y=\int_{-a}^{a}\sigma_y t\,dx=\int_{-a}^{a}(0)t\,dx=0$$ ◄

Edge y=b:

$$V_x=\int_{-a}^{a}\left(-\frac{3}{2}px^2b+c_1b^2+\frac{px^4}{2}+c_3\right)t\,dx$$
$$=-pa^3\left(b^2-\frac{a^2}{5}\right)t+2a(c_1b^2+c_3)t$$ ◄

$$P_y=\int_{-a}^{a}(pxb^3-2px^3b)t\,dx=0$$ ◄

3.2

Edge $x=\pm a$:

$\tau_{xy}=0$: $\quad -\frac{3}{2}pa^2y^2+c_1y^2+\frac{1}{2}pa^4+c_3$

$\tau_{xy}=0$: $\quad -\frac{3}{2}pa^2y^2+c_1y^2+\frac{1}{2}pa^4+c_3$

Adding,

$$(-3pa^2+2c_1)y^2+pa^4+2c_3=0$$

or

$$c_1=\frac{3}{2}pa^2 \qquad c_3=-\frac{1}{2}pa^4$$ ◄

Edge x=a:

$\sigma_x=0$: $\quad pa^3y-2c_1ay+c_2y=0$

or

$$c_2=2pa^3$$ ◄

3.3

It is readily shown that

$\nabla^4\Phi_1=0$ is satisfied

$\nabla^4\Phi_2=0$ is satisfied ◄

We have

$$\sigma_x=\frac{\partial^2\Phi_1}{\partial y^2}=2c,\quad \sigma_y=\frac{\partial^2\Phi_1}{\partial x^2}=2a,\quad \tau_{xy}=-\frac{\partial^2\Phi_1}{\partial x\partial y}=-b$$ ◄

Thus, stresses are uniform over the body.

Similarly, for Φ_2:

$$\sigma_x=2cx+6dy \qquad \sigma_y=6ax+2by$$
$$\tau_{xy}=-2bx-2cy$$ ◄

Thus, stresses vary linearly with respect to x and y over the body.

3.4

Note: Since $\sigma_z=0$ and $\varepsilon_y=0$, we have plane stress in xy plane and plane strain in xz plane, respectively.

Equations of compatibility and equilibrium are satisfied by

$$\sigma_x=-\sigma_o \qquad \sigma_y=-c \qquad \sigma_z=0 \tag{a}$$
$$\tau_{xy}=\tau_{yz}=\tau_{xz}=0$$

We have

$$\varepsilon_y=0 \tag{b}$$

Stress-strain relations become

$$\varepsilon_x=(\sigma_x-\nu\sigma_y)/E,\quad \varepsilon_y=(\sigma_y-\nu\sigma_x)/E \tag{c}$$
$$\varepsilon_z=-\nu(\sigma_x+\sigma_y)/E,\quad \gamma_{xy}=\gamma_{yz}=\gamma_{xz}=0$$

Substituting Eqs. (a,b) into Eqs. Eqs. (c), and solving

$$\sigma_y=-\nu\sigma_o \qquad \varepsilon_z=\nu(1+\nu)\sigma_o/E$$
$$\varepsilon_x=-(1-\nu^2)/\sigma_o E \qquad \varepsilon_y=0$$

Then, Eqs. (2.3) yield, after integrating:

$$u=-(1-\nu^2)\sigma_o x/E \qquad v=0$$
$$w=\nu(1+\nu)\sigma_o z/E$$ ◄

3.5

(a) $\sigma_x = \frac{\partial^2 \Phi}{\partial y^2} = 0, \quad \sigma_y = 6pxy, \quad \tau_{xy} = -3px^2$ ◄

Note that $\nabla^4 \Phi = 0$ is satisfied.

(b)

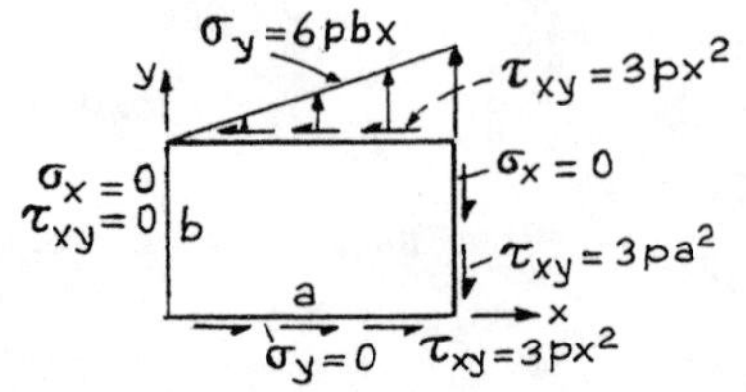

(c) Edge $x=0$: $\quad V_y = P_x = 0$ ◄

Edge $x=a$: $\quad P_x = 0$ ◄

$$V_y = \int_0^b \tau_{xy} t\, dy = 3pa^2 bt \downarrow$$

Edge $y=0$: $\quad P_y = 0$ ◄

$$V_x = \int_0^a \tau_{xy} t\, dx = pa^3 t \rightarrow$$

Edge $y=b$: $\quad V_x = pa^3 t \leftarrow$ ◄

$$P_y = \int_0^a \sigma_y t\, dx = 3pa^2 bt \uparrow$$

3.6

(a) We have $\nabla^4 \Phi = 0$ is satisfied.

$$\sigma_y = \frac{\partial^2 \Phi}{\partial x^2} = py^2/a^2, \quad \sigma_x = p(x^2+xy)/a^2$$ ◄

$$\tau_{xy} = -p(4xy+y^2)/2a^2$$

(b)

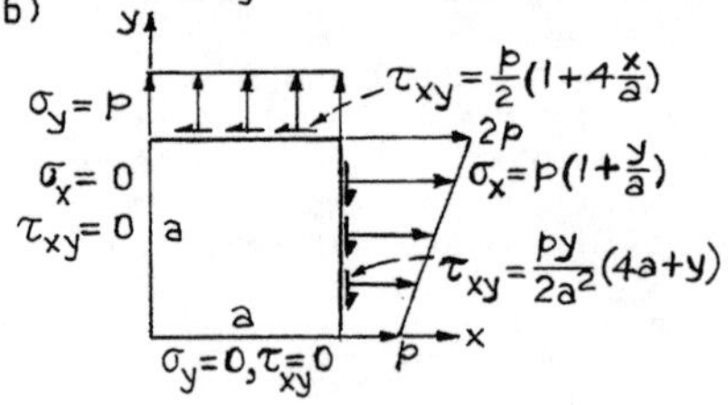

(c) Edge $x=0$: $\quad V_y = 0 \qquad P_x = 0$ ◄

Edge $x=a$:

$$V_y = \int_0^a \tau_{xy} t\, dy = \tfrac{7}{6} pat \downarrow$$ ◄

$$P_x = \int_0^a \sigma_x t\, dy = \tfrac{3}{2} pat \rightarrow$$

Edge $y=0$: $\quad V_x = 0 \qquad P_y = 0$ ◄

Edge $y=a$:

$$V_x = \int_0^a \tau_{xy} t\, dx = \tfrac{3}{2} pat \leftarrow$$ ◄

$$P_y = \int_0^a \sigma_y pt\, dx = pat \uparrow$$

3.7

We have

$$\frac{\partial \Phi}{\partial y} = -\frac{P}{\pi}\left[\tan^{-1}\frac{y}{x} + \frac{xy}{x^2+y^2}\right], \quad \frac{\partial \Phi}{\partial x} = -\frac{Py}{\pi}\frac{-y}{x^2+y^2}$$

$$\frac{\partial^2 \Phi}{\partial y^2} = -\frac{P}{\pi}\left[\frac{x}{x^2+y^2} + \frac{(x^2+y^2)x - 2y^2 x}{(x^2+y^2)^2}\right]$$

The stresses are thus,

$$\sigma_x = \frac{\partial^2 \Phi}{\partial y^2} = -\frac{2P}{\pi}\frac{x^3}{(x^2+y^2)^2}$$

$$\sigma_y = \frac{\partial^2 \Phi}{\partial x^2} = -\frac{2P}{\pi}\frac{xy^2}{(x^2+y^2)^2}$$ ◄

$$\tau_{xy} = -\frac{\partial^2 \Phi}{\partial x \partial y} = -\frac{2P}{\pi}\frac{x^2 y}{(x^2+y^2)^2}$$

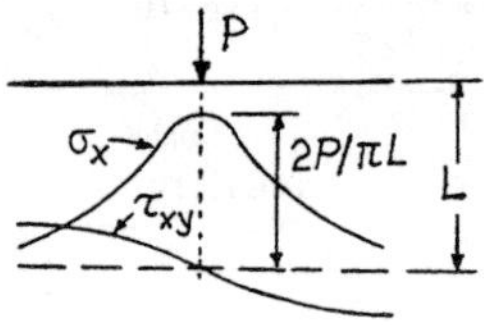

3.8

Various derivatives of Φ are:

$$\frac{\partial \Phi}{\partial x} = \frac{\tau_o}{4}\left(y - \frac{y^2}{h} - \frac{y^3}{h^2}\right), \quad \frac{\partial^2 \Phi}{\partial x^2} = 0$$

$$\frac{\partial^4 \Phi}{\partial x^2 \partial y^2} = 0, \qquad \frac{\partial^2 \Phi}{\partial x \partial y} = \frac{\tau_o}{4}\left(1 - \frac{2y}{h} - \frac{3y^2}{h^2}\right)$$

$$\frac{\partial^2 \Phi}{\partial y^2} = \frac{\tau_o}{4h}\left(-2x - \frac{6xy}{h} + 2L + \frac{6Ly}{h}\right) \qquad \text{(a)}$$

$$\frac{\partial^4 \Phi}{\partial x^4} = 0, \qquad \frac{\partial^4 \Phi}{\partial y^4} = 0$$

It is clear that Eqs. (a) satisfy Eq. (3.14). On the basis of Eq.(a) and (3.13), we obtain

$$\sigma_x = \frac{\tau_o}{4h}\left(-2x - \frac{6xy}{h} + 2L + \frac{6Ly}{h}\right), \quad \sigma_y = 0$$
$$\tau_{xy} = -\frac{\tau_o}{4}\left(1 - \frac{2y}{h} - \frac{3y^2}{h^2}\right) \qquad \text{(b)}$$

From Eqs. (b), we determine

Edge $y=h$: $\quad \sigma_y = 0 \qquad \tau_{xy} = \tau_o$

Edge $y=-h$: $\quad \sigma_y = 0 \qquad \tau_{xy} = 0$

Edge $x=L$:

$$\sigma_x = 0 \qquad \tau_{xy} = -\frac{\tau_o}{4}\left(1 - \frac{2y}{h} - \frac{3y^2}{h^2}\right)$$

It is observed from the above that boundary conditions are satisfied at $y = \pm h$, but not at $x=L$. ◄

3.9

We obtain

$$\sigma_x = \partial^2\Phi/\partial y^2 = p(x^2-2y^2)/a^2$$

$$\sigma_y = \partial^2\Phi/\partial x^2 = py^2/a^2 \qquad (a)$$

$$\tau_{xy} = -\partial^2\Phi/\partial x\partial y = -2pxy/a^2$$

Taking higher derivatives of Φ, it is seen that Eq. (3.14) is satisfied.

Stress field along the edges of the plate, as determined from Eqs. (a), is sketced below.

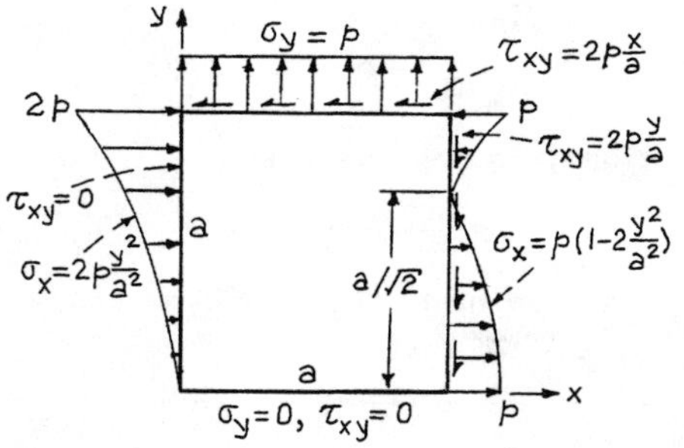

3.10

The first of Eqs. (3.6) with $F_x = 0$

$$\partial\tau_{xy}/\partial y = pxy/I$$

Integrating,

$$\tau_{xy} = \frac{pxy^2}{2I} + f_1(x) \qquad (a)$$

The boundary condition,

$$(\tau_{xy})_{y=h} = 0 = \frac{pxh^2}{2I} + f_1(x)$$

gives $f_1(x) = -pxh^2/2I$. Equation (a) becomes

$$\tau_{xy} = -\frac{px}{2I}(h^2-y^2) \qquad (b)$$

Clearly, $(\tau_{xy})_{y=-h} = 0$ is satisfied by Eq. (b).

Then, the second of Eqs. (3.6) with $F_y = 0$ results in

$$\partial\sigma_y/\partial y = p(h^2-y^2)/2I$$

(CONT.)

3.10 CONT.

Integrating,

$$\sigma_y = \frac{p}{2I}y\left(h^2-\frac{y^2}{3}\right) + f_2(x) \qquad (c)$$

Boundary conditions, with $t = 3I/2h^3$,

$$(\sigma_y)_{y=-h} = -\frac{p}{t} = -\frac{ph}{2I}\left(h^2-\frac{h^3}{3}\right) + f_2(x)$$

gives $f_2(x) = -ph^3/3I$. Equation (c) is thus

$$\sigma_y = \frac{p}{6I}(3h^2y - y^3 - 2h^3) \qquad (d)$$

This satisfies the condition that $(\sigma_y)_{y=h} = 0$.

Note that Eq. (3.12) is not satisfied: the solution obtained does not provide a compatible displacement field.

3.11

Substituting the stresses from Eqs. (3.10) into Eqs. (3.6) and taking $F_x = F_y = 0$:

$$\frac{E}{1-\nu^2}\left(\frac{\partial\varepsilon_x}{\partial x} + \nu\frac{\partial\varepsilon_y}{\partial x}\right) + G\frac{\partial\gamma_{xy}}{\partial y} = 0$$

$$\frac{E}{1-\nu^2}\left(\frac{\partial\varepsilon_y}{\partial y} + \nu\frac{\partial\varepsilon_x}{\partial y}\right) + G\frac{\partial\gamma_{xy}}{\partial x} = 0$$

or

$$\frac{E}{2(1+\nu)}\left[\frac{2}{1-\nu}\frac{\partial^2 v}{\partial x^2} + \frac{2\nu}{1-\nu}\frac{\partial^2 v}{\partial x\partial y} + \frac{\partial^2 u}{\partial y^2} + \frac{\partial^2 v}{\partial x\partial y}\right] = 0$$

$$\frac{E}{2(1+\nu)}\left[\frac{2}{1-\nu}\frac{\partial^2 v}{\partial y^2} + \frac{2\nu}{1-\nu}\frac{\partial^2 u}{\partial x\partial y} - \frac{\partial^2 u}{\partial x\partial y} + \frac{\partial^2 v}{\partial x^2}\right] = 0$$

The foregoing become

$$\frac{\partial^2 u}{\partial x^2} + \frac{1+\nu}{1-\nu}\frac{\partial^2 u}{\partial x^2} + \frac{1+\nu}{1-\nu}\frac{\partial^2 v}{\partial x\partial y} + \frac{\partial^2 u}{\partial y^2} = 0$$

$$\frac{\partial^2 v}{\partial y^2} + \frac{1+\nu}{1-\nu}\frac{\partial^2 v}{\partial y^2} + \frac{1+\nu}{1-\nu}\frac{\partial^2 u}{\partial x\partial y} + \frac{\partial^2 v}{\partial x^2} = 0$$

or

$$\frac{\partial^2 u}{\partial x^2} + \frac{\partial^2 u}{\partial y^2} + \frac{1+\nu}{1-\nu}\frac{\partial}{\partial x}\left(\frac{\partial u}{\partial x} + \frac{\partial v}{\partial y}\right) = 0$$

$$\frac{\partial^2 v}{\partial y^2} + \frac{\partial^2 v}{\partial x^2} + \frac{1+\nu}{1-\nu}\frac{\partial}{\partial y}\left(\frac{\partial v}{\partial y} + \frac{\partial u}{\partial x}\right) = 0$$

3.12

It is readily found that

$$\partial\sigma_x/\partial x=-pxy/I \qquad \partial\tau_{xy}/\partial y=pxy/I$$

and

$$\partial\sigma_y=-\frac{p}{I}(-h^2+y^2) \qquad \frac{\partial\tau_{xy}}{\partial x}=-\frac{p}{2I}(h^2-y^2)$$

Thus, Eqs. (3.6) are satisfied: stress field is possible.

We have x_Q=1.5 m, y_Q=0.05 m, and

$$I=2th^3/3=2(0.04)(0.1)^3/3$$
$$2.67(10^{-5})\ m^4$$

Substituting the given data, Eqs. (P3.12) yield at Q:

$$\sigma_x=\frac{-10\times10^3}{10(2.67\times10^{-5})}(5\text{x}0.5^2+2\text{x}0.1^2)0.05+\frac{-10\times10^3}{3(2.67\times10^{-5})}(1.25\text{x}10^{-4})$$
$$=-21.09\ \text{MPa}$$

$$\tau_{xy}=\frac{-10\times10^3}{2(2.67\times10^{-5})}(0.1^2-0.05^2)$$
$$=-2.106\ \text{MPa}$$

$$\sigma_y=\frac{-10\times10^3}{6(2.67\times10^{-5})}(2\text{x}0.1^3-3\text{x}0.0005+0.05^3)$$
$$=-0.039\ \text{MPa}$$

Applying Eq. (2.29), we have

$$G=\frac{200(10^9)}{2(1+0.3)}=76.9\ \text{MPa}$$

Hooke's law is therefore

$$\varepsilon_x=\frac{1}{200(10^9)}(-21.09+0.3\text{x}0.039)10^6$$
$$=-105\mu$$

$$\varepsilon_y=\frac{1}{200(10^9)}(-0.039+0.3\text{x}21.09)10^6$$
$$=31.4\mu$$

$$\gamma_{xy}=\frac{-2.106(10^6)}{76.9(10^9)}=-27.4\mu$$

Principal strains are

$$\varepsilon_{1,2}=\frac{-105+31.4}{2}\pm[(\frac{136.4}{2})^2+(\frac{-27.4}{2})^2]^{1/2}$$

or

$$\varepsilon_1=32.8\mu \qquad \varepsilon_2=-106\mu$$ ◄

We have

$$\theta_p=\frac{1}{2}\tan^{-1}\frac{-27.4}{-105-31.4}=5.65°$$

For this angle, Eq. (2.11a) yield $\varepsilon_{x'}$=32.8μ. Thus,

$$\theta_p'=5.65°$$ ◄

3.13

Assume

$\varepsilon_x=\varepsilon_y=0, \quad \sigma_z=0, \quad \sigma_x=\sigma_y=\text{constant}$

which satisfy Eqs. (3.6) and (3.24). Hooke's law becomes

$$\varepsilon_x=\frac{1}{E}(\sigma_x-\nu\sigma_y)+\alpha T_1=0 \qquad \text{(a)}$$

$$\varepsilon_y=\frac{1}{E}(\sigma_y-\nu\sigma_x)+\alpha T_1=0 \qquad \text{(b)}$$

$$\varepsilon_z=\frac{\nu}{E}(-\sigma_x-\sigma_y)+\alpha T_1=0 \qquad \text{(c)}$$

From Eqs. (a) and (b), we obtain $\sigma_x=\sigma_y$. Therefore,

$$\varepsilon_x=\frac{\sigma_x}{E}(1-\nu)+\alpha T_1=0$$ ◄

This yields

$$\sigma_x=\sigma_y=\frac{E\alpha T_1}{\nu-1}$$ ◄

Then, Eq. (c) becomes

$$\varepsilon_z=\frac{2\nu\alpha T_1}{1-\nu}+\alpha T_1$$ ◄

We also have: $\tau_{xy}=\tau_{yz}=\tau_{xz}=0$ and $\gamma_{xy}=\gamma_{yz}=\gamma_{xz}$.

3.14

The nonzero strain components are

$$\varepsilon_x=\varepsilon_y=\varepsilon_z=\alpha T$$

Compatibility equations reduce to

$$\frac{\partial^2\varepsilon_x}{\partial y^2}+\frac{\partial^2\varepsilon_y}{\partial x^2}=\frac{\partial^2\gamma_{xy}}{\partial y\partial x}$$

$$\frac{\partial^2\varepsilon_y}{\partial z^2}+\frac{\partial^2\varepsilon_z}{\partial y^2}=\frac{\partial^2\gamma_{yz}}{\partial y\partial z}$$

$$\frac{\partial^2\varepsilon_z}{\partial x^2}+\frac{\partial^2\varepsilon_x}{\partial z^2}=\frac{\partial^2\gamma_{xz}}{\partial x\partial z}$$

Adding the above equations and substituting the given data:

$$2\alpha(\frac{\partial^2T}{\partial x^2}+\frac{\partial^2T}{\partial y^2}+\frac{\partial^2T}{\partial z^2})=0$$

or

$$\frac{\partial^2T}{\partial x^2}+\frac{\partial^2T}{\partial y^2}+\frac{\partial^2T}{\partial z^2}=0$$

This equation, for a time independent temperature field, has the solution

$$T=c_1'x+c_2'y+c_3'z+c_4'$$

which may be written as

$$\alpha T=c_1x+c_2y+c_3z+c_4$$ ◄

3.15

Stress is $\sigma_x = -\sigma_o$ regardless of T_1 and still we have $\varepsilon_y = 0$. The second of Eqs. (3.23a) is thus

$$\varepsilon_y = \frac{1}{E}(\sigma_y + \nu\sigma_o) + \alpha T_1 = 0$$

or

$$\sigma_y = -\nu\sigma_o - E\alpha T_1 \qquad \text{(a)}$$

The first of Eqs. (3.23a) and (a) result in

$$\varepsilon_x = -\frac{1-\nu^2}{E}\sigma_o^2 + (1+\nu)\alpha T_1 \qquad \text{(b)}$$

Now, Hooke's law

$$\varepsilon_z = -\frac{\nu}{E}(\sigma_x - \sigma_y) + \alpha T_1$$

leads to

$$\varepsilon_z = \frac{\nu(1+\nu)}{E}\sigma_o + (1+\nu)\alpha T_1 \qquad \text{(c)}$$

Then Equations (2.3) yield, after integration,

$$u = \varepsilon_x x \qquad v = 0 \qquad w = \varepsilon_z z \quad ◄$$

Here, ε_x and ε_z are given by Eqs. (b) and (c).

3.16

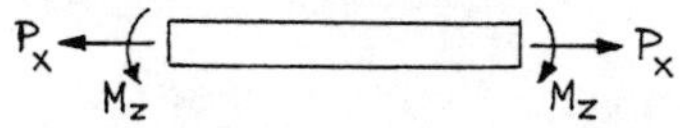

Equation (3.24) reduces to

$$\frac{d^2}{dy^2}(\sigma_x + \alpha ET) = 0$$

from which

$$\sigma_x = -\alpha ET + c_1 y + c_2$$
$$= -\alpha ET(a_1 y + a_2) + c_1 y + c_2$$

Referring to Part (a) of Example 3.2: $c_1 = c_2 = 0$ and

$$\sigma_x = -\alpha E(a_1 y + a_2)$$

We have

$$P_x = \int_{-h}^{h} \sigma_x t\, dy = -2E\alpha h t a_2 \quad ◄$$

$$M_z = \int_{-h}^{h} \sigma_x t y\, dy = -E\alpha t \left| \frac{a_1 y^3}{3} + \frac{a_2 y^2}{2} \right|_{-h}^{h}$$
$$= -\frac{2}{3}E\alpha t h^3 a_1 \quad ◄$$

The P_x and M_z are opposite to that shown above.

3.17

Assume a stress distribution:

$$\tau_{xy} = \tau_{yz} = \tau_{xz} = \sigma_y = \sigma_z = 0$$
$$\sigma_x = \text{constant} \qquad \text{(a)}$$

which satisfy Eqs. (3.6) and (3.24). Then, Hooke's law becomes

$$\varepsilon_x = \frac{\sigma_x}{E} + \alpha T, \quad \varepsilon_y = \varepsilon_z = -\frac{\nu\sigma_x}{E} + \alpha T \qquad \text{(b,c)}$$

$$\gamma_{xy} = \gamma_{yz} = \gamma_{xz} = 0 \qquad \text{(d)}$$

Due to constraint imposed by the walls and because of the uniformity of the temperature distribution: we take $\varepsilon_x = 0$. Equations (a) and (b) give

$$\sigma_x = -\alpha ET$$

Equation (c) is then

$$\varepsilon_y = \varepsilon_z = \alpha T(1+\nu) = \text{constant}$$

The compressive force P on the tube is

$$P_x = \sigma_x A = -E\alpha AT \quad ◄$$

Substituting the data given

$$P_x = -120(10^9)(16.8\times10^{-6})\times(800\times10^{-6})(100)$$
$$= -161.3 \text{ kN} \quad ◄$$

3.18

Derivatives of the given stress function are

$$\frac{\partial^2\Phi}{\partial r^2} = 0, \qquad \frac{1}{r}\frac{\partial\Phi}{\partial r} = -\frac{P\theta}{\pi r}\sin\theta$$

$$\frac{1}{r^2}\frac{\partial^2\Phi}{\partial\theta^2} = \frac{P\theta}{\pi r}\sin\theta - \frac{2P}{\pi r}\cos\theta$$

Equation (3.37) beomes

$$\nabla^4\Phi = \left(\frac{\partial^2}{\partial r^2} + \frac{1}{r}\frac{\partial}{\partial r} + \frac{1}{r^2}\frac{\partial^2}{\partial\theta^2}\right)\left(-\frac{2P}{\pi r}\cos\theta\right)$$

After performing the derivatives, we obtain

$$\nabla^4\Phi = 0$$

Substituting the derivatives obtained above, into Eqs. (3.29):

$$\sigma_r = -\frac{P\theta}{\pi r}\sin\theta + \frac{P\theta}{\pi r}\sin\theta - \frac{2P}{\pi r}\cos\theta$$
$$= -\frac{2P}{\pi r}\cos\theta \quad ◄$$

Similarly, we find $\sigma_\theta = 0$ and $\tau_{r\theta} = 0$.

3.19

Refer to Fig. P3.19a. Let $A_{BC}=1$ and hence $A_{AB}=\cos\theta$, $A_{AC}=\sin\theta$.

$\Sigma F_x=0$:

$$\sigma_x=\sigma_r\cos\theta\cos\theta+\sigma_\theta\sin\theta\sin\theta-2\tau_{r\theta}\sin\theta\cos\theta$$

$\Sigma F_y=0$:

$$\tau_{xy}=\sigma_r\cos\theta\sin\theta-\sigma_\theta\sin\theta\cos\theta+\tau_{r\theta}\cos\theta\cos\theta-\tau_{r\theta}\sin\theta\sin\theta$$

Similarly, from Fig. P3.19b:

$\Sigma F_y=0$:

$$\sigma_y=\sigma_r\sin^2\theta+\sigma_\theta\cos^2\theta+2\tau_{r\theta}\sin\theta\cos\theta$$

Check:

$\Sigma F_x=0$:

$$\tau_{xy}=\sigma_r\sin\theta\cos\theta-\sigma_\theta\sin\theta\cos\theta+\tau_{r\theta}(\cos^2\theta-\sin^2\theta)$$

Thus, quoted equations are derived.

3.20

Apply the chain rule (Sec. 3.7):

$$\frac{\partial\Phi}{\partial x}=\frac{\partial\Phi}{\partial r}\cos\theta-\frac{1}{r}\frac{\partial\Phi}{\partial\theta}\sin\theta$$

and

$$\frac{\partial^2\Phi}{\partial x^2}=\frac{\partial^2\Phi}{\partial r^2}\cos^2\theta-2\frac{\partial^2\Phi}{\partial\theta\partial r}\frac{\sin\theta\cos\theta}{r}+\frac{\partial\Phi}{\partial r}\frac{\sin^2\theta}{r}-2\frac{\partial\Phi}{\partial r}\frac{\sin\theta\cos\theta}{r^2}+\frac{\partial^2\Phi}{\partial\theta^2}\frac{\sin^2\theta}{r^2}\qquad\text{(a)}$$

Similarly,

$$\frac{\partial^2\Phi}{\partial y^2}=\frac{\partial^2\Phi}{\partial r^2}\sin^2\theta+2\frac{\partial^2\Phi}{\partial\theta\partial r}\frac{\sin\theta\cos\theta}{r}+\frac{\partial\Phi}{\partial r}\frac{\cos^2\theta}{r}+2\frac{\partial\Phi}{\partial\theta}\frac{\sin\theta\cos\theta}{r^2}+\frac{\partial^2\Phi}{\partial\theta^2}\frac{\cos^2\theta}{r^2}\qquad\text{(b)}$$

Adding Eqs. (a) and (b), we have

$$\frac{\partial^2\Phi}{\partial x^2}+\frac{\partial^2\Phi}{\partial y^2}=\frac{\partial^2\Phi}{\partial r^2}+\frac{1}{r}\frac{\partial\Phi}{\partial r}+\frac{1}{r^2}\frac{\partial^2\Phi}{\partial\theta^2}\qquad\text{(c)}$$

By referring to the identity

$$\frac{\partial^4\Phi}{\partial x^4}+2\frac{\partial^4\Phi}{\partial x^2\partial y^2}+\frac{\partial^4\Phi}{\partial y^4}=\left(\frac{\partial^2}{\partial x^2}+\frac{\partial^2}{\partial y^2}\right)\left(\frac{\partial^2\Phi}{\partial x^2}+\frac{\partial^2\Phi}{\partial y^2}\right)$$

and Eq. (c), we can readily write the equation quoted, Eq. (3.37).

3.21

Equation (3.25) is written as

$$\left(\frac{d^2}{dr^2}+\frac{1}{r}\frac{d}{dr}\right)\left(\frac{d^2\Phi}{dr^2}+\frac{1}{r}\frac{d\Phi}{dr}\right)+\alpha ET\left(\frac{d^2}{dr^2}+\frac{1}{r}\frac{d}{dr}\right)=0$$

or

$$\frac{d^2\Phi}{dr^2}+\frac{1}{r}\frac{d\Phi}{dr}+\alpha ET=0$$

or

$$\frac{1}{r}\frac{d}{dr}\left(r\frac{d\Phi}{dr}\right)+\alpha ET=0 \quad ◄$$

3.22

(a) Let $C=-M/2(\sin2\alpha-2\alpha\cos2\alpha)$, and
$\Phi=C(\sin2\theta-2\theta\cos2\alpha)$

Various derivatives of Φ are:

$\partial\Phi/\partial r=0 \qquad \partial^2\Phi/\partial r^2=0$

$\partial\Phi/\partial\theta=2C\cos2\theta-2C\cos2\alpha$

$\partial^2\Phi/\partial\theta^2=-4C\sin2\theta$

and

$$\nabla^2\Phi=\frac{1}{r^2}\frac{\partial^2\Phi}{\partial\theta^2}=-4C\sin2\theta/r^2$$

We thus obtain

$$\nabla^4\Phi=C\left[\frac{8\sin2\theta}{r^4}-\frac{24\sin2\theta}{r^4}+\frac{16\sin2\theta}{r^4}\right]=0 \quad ◄$$

(b) $$\sigma_r=\frac{1}{r^2}\frac{\partial^2\Phi}{\partial\theta^2}=-\frac{4C\sin2\theta}{r^2}$$

$$\sigma_\theta=0 \quad ◄$$

$$\tau_{r\theta}=-\frac{\partial}{\partial r}\left(\frac{1}{r}\frac{\partial\Phi}{\partial\theta}\right)=-\frac{2C}{r^2}(\cos2\theta-\cos2\alpha)$$

(c) Letting $\alpha=\pi/2$: $C=-M/2\pi$. It follows that

$$\sigma_r=\frac{2M\sin2\theta}{\pi r^2}\qquad\sigma_\theta=0$$

$$\tau_{r\theta}=\frac{2C}{\pi r^2}(\cos2\theta+1)=\frac{2M\cos^2\theta}{\pi r^2} \quad ◄$$

where $\cos^2\theta=(1+\cos2\theta)/2$.

3.23

Using Eq. (3.45) and Fig. P3.23:

$$F_x=\int_0^{\pi/2}(\sigma_r r\,d\theta)\sin\theta=\int_0^{\pi/2}\left(\frac{2P}{\pi}\cos\theta\sin\theta\right)d\theta$$

$$=\frac{2P}{\pi}\left|\frac{1}{2}\sin^2\theta\right|_0^{\pi/2}=\frac{P}{\pi}\ \rightleftarrows \quad ◄$$

Similarly,

$$F_y=\int_{-\pi/2}^{\pi/2}(\sigma_r r\,d\theta)\cos\theta=\int_{-\pi/2}^{\pi/2}\left(\frac{2P}{\pi}\cos^2\theta\right)d\theta$$

$$=\frac{2P}{\pi}\left|\frac{\theta}{2}+\frac{1}{4}\sin2\theta\right|_{-\pi/2}^{\pi/2}=P\downarrow \quad ◄$$

3.24

NOTE: (In Probs. 3.24, 3.25, and 3.26), the P, L, and α are constants. It can readily be verified that, the maximum values of the functions in parantheses occurs:

$$\frac{d}{d\theta}(\sin\theta\cos^3\theta)=0, \text{ or } \tan^2\theta=1/3 \text{ when } \theta=\pm 30^\circ$$

$$\frac{d}{d\theta}(\sin^2\theta\cos^2\theta)=0, \text{ or } \tan^2\theta=1 \text{ when } \theta=\pm 45^\circ$$

Maximum stresses, using Eqs. (3.34) and (3.40):

$$(\sigma_x)_{elast.}=\frac{P}{L(\alpha+\frac{1}{2}\sin 2\alpha)}, \quad (\tau_{xy})_{elast.}=\frac{P\sin\theta\cos^3\theta}{L(\alpha+\frac{1}{2}\sin 2\alpha)} \qquad \text{(a)}$$

Elementary solution of maximum stresses are

$$(\sigma_x)_{elem.}=\frac{P}{2L\tan\alpha}, \quad (\tau_{xy})_{elem.}=0 \qquad \text{(b)}$$

(a) For $\alpha=15^\circ$, $(\alpha+\frac{1}{2}\sin 2\alpha=0.512)$:

$(\sigma_x)_{elast.}=P/0.512L$ at $\theta=0^\circ$
$(\tau_{xy})_{elast.}=P/2.195L$ at $\theta=15^\circ$ ◄
$(\sigma_x)_{elem.}=P/0.536L$ at any θ

Thus, $(\sigma_x)_{elast.}=1.047(\sigma_x)_{elem.}$

(b) For $\alpha=60^\circ$, $(\alpha+\frac{1}{2}\sin 2\alpha=1.48)$:

$(\sigma_x)_{elast.}=P/1.48L$ at $\theta=0^\circ$
$(\tau_{xy})_{elast.}=P/4.557L$ at $\theta=30^\circ$
$(\sigma_x)_{elem.}=P/3.464L$ at any θ
Thus, $(\sigma_x)_{elast.}=2.341(\sigma_x)_{elem.}$ ◄

3.25

See: **NOTE**, solution of Prob. 3.24.

Substitute $\alpha=30^\circ$ into Eqs. (a) and (b) of Solution of Prob. 3.24.

$(\sigma_x)_{elast.}=P/0.957L$ at $\theta=0^\circ$
$(\tau_{xy})_{elast.}=P/2.946L$ at $\theta=30^\circ$
$(\sigma_x)_{elem.}=P/1.155L$ at any θ

Thus,

$(\sigma_x)_{elast.}=1.207(\sigma_x)_{elem.}$ ◄

3.26

See: **NOTE**, solution of Prob. 3.24.

We have $r=L/\cos\theta$, $h_{mn}/2=c=L\cdot\tan\alpha$, and $I=2c^3/3$. Equations (3.43) give

$$(\sigma_x)_{elast.}=\frac{F\sin\theta\cos^3\theta}{L(\alpha-\frac{1}{2}\sin 2\alpha)}$$

$$(\tau_{xy})_{elast.}=\frac{F\sin^2\theta\cos^2\theta}{L(\alpha-\frac{1}{2}\sin 2\alpha)}$$ ◄

Elementary solution:

$(\sigma_x)_{elem.}=3FL/2c^2$, $(\tau_{xy})_{elem.}=3F/4c$ ◄

(a) For $\alpha=15^\circ$, $(\alpha-\frac{1}{2}\sin 2\alpha=0.012)$:

$(\sigma_x)_{elast.}=19.43F/L$ at $\theta=15^\circ$
$(\tau_{xy})_{elast.}=5.21F/L$ at $\theta=15^\circ$
$(\sigma_x)_{elem.}=20.89F/L$ at $\theta=15^\circ$
$(\tau_{xy})_{elem.}=2.8F/L$ at $\theta=0^\circ$

Thus, $(\sigma_x)_{elast.}=0.93(\sigma_x)_{elem.}$
$(\tau_{xy})_{elast.}=1.86(\tau_{xy})_{elem.}$ ◄

(b) For $\alpha=60^\circ$, $(\alpha-\frac{1}{2}\sin 2\alpha=0.614)$:

$(\sigma_x)_{elast.}=0.529F/L$ at $\theta=30^\circ$
$(\tau_{xy})_{elast.}=0.407F/L$ at $\theta=45^\circ$
$(\sigma_x)_{elem.}=0.5F/L$ at $\theta=60^\circ$
$(\tau_{xy})_{elem.}=0.433F/L$ at $\theta=0^\circ$

Thus, $(\sigma_x)_{elast.}=1.058(\sigma_x)_{elem.}$
$(\tau_{xy})_{elast.}=0.94(\tau_{xy})_{elem.}$ ◄

3.27

With $x=r\cdot\cos\theta$, Eqs. (3. 47) become

$$\sigma_x=-(2P/\pi r)\cos^3\theta$$
$$\sigma_y=-(2P/\pi r)\sin^2\theta\cos\theta$$
$$\tau_{xy}=-(2P/\pi r)\sin\theta\cos^2\theta$$

Substituting for P:

$$d\sigma_x=-\frac{2}{\pi r}\left(\frac{p\,r\,d\theta}{\cos\theta}\right)\cos^3\theta=-\frac{2p}{\pi}\cos^2\theta\,d\theta$$

$$d\sigma_y=-\frac{2p}{\pi}\sin^2\theta, \quad d\tau_{xy}=-\frac{p}{\pi}\sin 2\theta\,d\theta$$

Integrating,

$$\sigma_x=-\frac{2p}{\pi}\int_{\theta_1}^{\theta_2}\cos^2\theta\,d\theta$$
$$=-\frac{p}{2\pi}[2(\theta_2-\theta_1)+(\sin 2\theta_2-\sin 2\theta_1)]$$

$$\sigma_y=-\frac{p}{2\pi}[2(\theta_2-\theta_1)-(\sin 2\theta_2-\sin 2\theta_1)]$$ ◄

$$\tau_{xy}=(p/2\pi)[\cos 2\theta_2-\cos 2\theta_1]$$

3.28

APPROACH (a):

$$\left(\frac{d^2}{dr^2}+\frac{1}{r}\frac{d}{dr}\right)\left(\frac{d^2f_1}{dr^2}+\frac{1}{r}\frac{df_1}{dr}\right)=0 \qquad \text{(d)}$$

We have

$$\frac{d}{dr}\left(\frac{d^2f_1}{dr^2}\right)=\frac{d^3f_1}{dr^3}, \quad \frac{d^2}{dr^2}\left(\frac{d^2f_1}{dr^2}\right)=\frac{d^4f_1}{dr^4}$$

$$\frac{d}{dr}\left(\frac{1}{r}\frac{df_1}{dr}\right)=-r^{-2}\frac{df_1}{dr}+r^{-1}\frac{d^2f_1}{dr^2}$$

$$\frac{d^2}{dr^2}\left(\frac{1}{r}\frac{df_1}{dr}\right)=\frac{2}{r^3}\frac{df_1}{dr}-\frac{2}{r}\frac{d^2f_1}{dr^2}+\frac{1}{r}\frac{d^3f_1}{dr^3}$$

$$\frac{1}{r}\frac{d}{dr}\left(\frac{d^2f_1}{dr^2}\right)=\frac{1}{r}\frac{d^3f_1}{dr^3}$$

$$\frac{1}{r}\frac{d}{dr}\left(\frac{1}{r}\frac{df_1}{dr}\right)=\frac{1}{r}\left(-\frac{1}{r^2}\frac{df_1}{dr}+\frac{1}{r}\frac{d^2f_1}{dr^2}\right)$$

Then, Eq. (d) becomes

$$\frac{d^4f_1}{dr^4}+\frac{2}{r}\frac{d^3f_1}{dr^3}-\frac{1}{r^2}\frac{d^2f_1}{dr^2}+\frac{1}{r^3}\frac{df_1}{dr}=0 \qquad \text{(d')}$$

The first equation of Problem 3.28 may be written as:

$$\frac{1}{r}\frac{d}{dr}\{r\frac{d}{dr}[r^{-1}\frac{d}{dr}(r\frac{df_1}{dr})]\}=0$$

$$\frac{1}{r}\frac{d}{dr}\{r\frac{d}{dr}[r^{-1}\frac{df_1}{dr}+\frac{d^2f_1}{dr^2}]\}=0$$

$$\frac{1}{r}\frac{d}{dr}\{r[-r^{-2}\frac{df_1}{dr}+r^{-1}\frac{d^2f_1}{dr^2}+\frac{d^3f_1}{dr^3}]\}=0$$

$$\frac{1}{r}\frac{d}{dr}\{-r^{-1}\frac{df_1}{dr}+\frac{d^2f_1}{dr^2}+r\frac{d^3f_1}{dr^3}\}=0$$

$$\frac{1}{r}\{r^{-2}\frac{df_1}{dr}-r^{-1}\frac{d^2f_1}{dr^2}+2\frac{d^3f_1}{dr^3}+r\frac{d^4f_1}{dr^4}\}=0$$

or

$$\frac{1}{r^3}\frac{df_1}{dr}-\frac{1}{r^2}\frac{d^2f_1}{dr^2}+\frac{2}{r}\frac{d^3f_1}{dr^3}+\frac{d^4f_1}{dr^4}=0$$

which is the same as Eq. (d').

Now let us integrate the expression:

$$\frac{1}{r}\frac{d}{dr}\{r\frac{d}{dr}[\frac{1}{r}\frac{d}{dr}(r\frac{df_1}{dr})]\}=0$$

$$r\frac{d}{dr}[\frac{1}{r}\frac{d}{dr}(r\frac{df_1}{dr})]=c_1$$

$$\frac{1}{r}\frac{d}{dr}(r\frac{df_1}{dr})=c_1\ln r+c_2$$

(CONT.)

3.28 CONT.

$$r\frac{df_1}{dr}=c_1\int r\ln r\,dr+c_2\int r\,dr$$

$$r\frac{df_1}{dr}=c_1\left[\frac{r^2}{2}\ln r-\frac{r^4}{4}\right]+c_2\frac{r^2}{2}+c_3$$

$$\frac{df_1}{dr}=c_1 r\ln r+c_2 r+\frac{c_3}{r}$$

or

$$f_1=c_1 r^2\ln r+c_2 r^2+c_3\ln r+c_4$$

Expression (e) may be treated in a like manner.

APPROACH (b):

Letting $t=\ln r$, we have

$$\frac{df_1}{dr}=\frac{df_1}{dt}\frac{dt}{dr}=\frac{1}{r}\frac{df_1}{dr}$$

$$\frac{d^2f_1}{dr^2}=\frac{1}{r^2}\left(\frac{d^2f_1}{dt^2}-\frac{df_1}{dt}\right)$$

$$\frac{d^3f_1}{dr^3}=\frac{1}{r^3}\left(\frac{d^3f_1}{dt^3}-3\frac{d^2f_1}{dt^2}+2\frac{df_1}{dt}\right)$$

$$\frac{d^4f_1}{dr^4}=\frac{1}{r}\left(\frac{d^4f_1}{dt^4}-6\frac{d^3f_1}{dt^3}+11\frac{d^2f_1}{dt^2}-6\frac{df_1}{dt}\right)$$

Substituting these derivatives into Eq. (d'), we obtain:

$$\frac{d^4f_1}{dt^4}-4\frac{d^3f_1}{dt^3}+4\frac{d^2f_1}{dt^2}=0$$

This is an ordinary differential equation with constant coefficients. It has a solution

$$f_1=c_1 r^2\ln r+c_2 r^2+c_3\ln r+c_4$$

In a like manner, it can be shown that,

$$\left(\frac{d^2}{dr^2}+\frac{1}{r}\frac{d}{dr}-\frac{4}{r^2}\right)\left(\frac{d^2f_2}{dr^2}+\frac{1}{r}\frac{df_2}{dr^2}-\frac{4f_2}{r^2}\right)=0$$

is solved to yield Eq. (g) of Section 3.10.

3.29

(a)

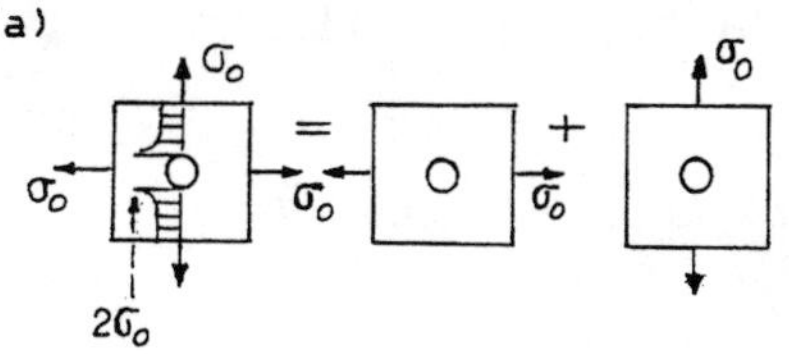

$$\sigma_{r1}=\frac{\sigma_o}{2}\left[\left(1-\frac{a^2}{r^2}\right)+\left(1+\frac{3a^4}{r^4}-\frac{4a^2}{r^2}\right)\cos 2\theta\right]$$

$$\sigma_{\theta 1}=\frac{\sigma_o}{2}\left[\left(1+\frac{a^2}{r^2}\right)-\left(1+\frac{3a^4}{r^4}\right)\cos 2\theta\right]$$

$$\tau_{r\theta 1}=-\frac{\sigma_o}{2}\left(1-\frac{3a^4}{r^4}+\frac{2a^2}{r^2}\right)\sin 2\theta$$

and

$$\sigma_{r2}=\frac{\sigma_o}{2}\left[\left(1-\frac{a^2}{r^2}\right)+\left(1+\frac{3a^4}{r^4}-\frac{4a^2}{r^2}\right)\times \cos 2(\theta+90^\circ)\right]$$

$$\sigma_{\theta 2}=\frac{\sigma_o}{2}\left[\left(1+\frac{a^2}{r^2}\right)-\left(1+\frac{3a^4}{r^4}\right)\times \cos 2(\theta+90^\circ)\right]$$

$$\tau_{r\theta 2}=-\frac{\sigma_o}{2}\left(1-\frac{3a^4}{r^4}+\frac{2a^2}{r^2}\right)\cos 2(\theta+90^\circ)$$

We have, by superposition:

$$\sigma_r=\sigma_{r1}+\sigma_{r2} \qquad \sigma_\theta=\sigma_{\theta 1}+\sigma_{\theta 2}$$

$$\tau_{r\theta}=\tau_{r\theta 1}+\tau_{r\theta 2}$$

Hence, at $r=a$ and $\theta=\pi/2$,

$$\sigma_{r1}=0 \qquad \sigma_{r2}=0$$

$$\sigma_{\theta 1}=3\sigma_o \qquad \sigma_{\theta 2}=-\sigma_o$$

$$\tau_{r\theta 1}=0 \qquad \tau_{r\theta 2}=0$$

lead to the solution:

$$\sigma_r=0 \qquad \sigma_\theta=2\sigma_o \qquad \tau_{r\theta}=0$$ ◄

(b) Referring to the results of part (a), we write

$$\sigma_{r1}=0 \qquad \sigma_{r2}=0$$

$$\sigma_{\theta 1}=3\sigma_o \qquad \sigma_{\theta 2}=\sigma_o$$

$$\tau_{r\theta 1}=0 \qquad \tau_{r\theta 2}=0$$

Thus,

$$\sigma_r=0 \qquad \sigma_\theta=4\sigma_o \qquad \tau_{r\theta}=0$$ ◄

3.30

We have

$\sqrt{h/a}=1 \qquad \sqrt{b/a}=4.123$

From curve 5, scale f:

$k\approx 3$ ◄

3.31

(a) We have

$\sqrt{h/a}=1.732 \qquad \sqrt{b/a}=3.873$

Using curve 1, scale f:

$k\approx 3.65$ ◄

(b) Curve 2, scale f:

$k\approx 3$ ◄

3.32

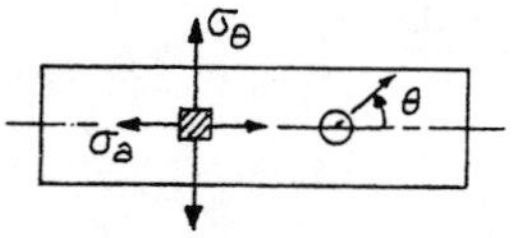

Without hole:

$\sigma_\theta=pd/2t \qquad \sigma_a=pd/4t$

With hole:

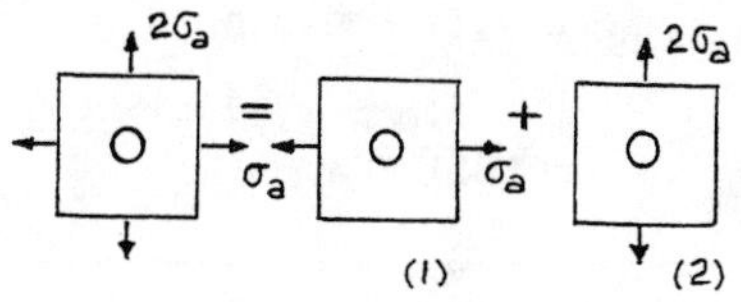

We use Eq. (3.52b), with $r=a$.

$$\sigma_{\theta 1}=\frac{\sigma_a}{2}\left[\left(1+\frac{a^2}{a^2}\right)-\left(1+\frac{3a^4}{a^4}\right)\cos 2\theta\right]$$

$$=\sigma_a[1-2\cos 2\theta]$$ ◄

$$\sigma_{\theta 2}=2\sigma_a[1-2\cos 2(\theta+90^\circ)]$$ ◄

For $\theta=0^\circ$:

$$\sigma_{\theta 1}=-\sigma_a \qquad \sigma_{\theta 2}=6\sigma_a$$

For $\theta=\pi/2$:

$$\sigma_{\theta 1}=3\sigma_a \qquad \sigma_{\theta 2}=-2\sigma_a$$

Therefore, superposing the results at $\theta=0^\circ$:

$$\sigma_\theta=5\sigma_a=5pd/4t$$ ◄

at $\theta=\pi/2$:

$$\sigma_\theta=\sigma_a=pd/4t$$ ◄

3.33

(a) We find from Table 3.2 that

$k_t = 2.10 \qquad k_b = 3.15 \qquad k_a = 3.40$

Then, Eqs. (j) of Example 3.5 yield

$$\sigma_x = 3.40\frac{10(10^3)}{\pi(0.2)^2} + 3.15\frac{4(2\times10^3)}{\pi(0.2)^3}$$

$$= 1.273 \text{ MPa}$$

$$\tau_{xy} = 2.10\frac{2(4\times10^3)}{\pi(0.2)^3} = 0.668 \text{ MPa}$$

Equation (i) of Example 3.5 is therefore

$$\sigma_{1,2} = \frac{1.273}{2} \pm [(\frac{1.273}{2})^2 + (0.668)^2]^{1/2}$$

or

$$\sigma_1 = 1.559 \text{ MPa}$$

$$\sigma_2 = -0.286 \text{ MPa}$$ ◄

(b) $\tau_{max} = \frac{1}{2}(1.559 + 0.286)$

$= 0.923$ MPa ◄

(c) $\sigma_{oct.} = \frac{1}{3}(1.559 - 0.286)$

$= 0.424$ MPa ◄

$$\tau_{oct.} = \frac{1}{3}[(1.559+0.286)^2 + (-0.286)^2 + (-1.559)^2]^{1/2}$$

$= 0.811$ MPa ◄

3.34

(a) We find from Table 3.2:

$k_b = 2.90 \qquad k_t = 1.95$

Equation (j) of Example 3.5 is therefore

$$\sigma_x = 2.90\frac{4(1.5\times10^3)}{\pi(0.12)^3} = 3.205 \text{ MPa}$$

$$\tau_{xy} = 1.95\frac{2(3\times10^3)}{\pi(0.12)^3} = 2.155 \text{ MPa}$$

Equation (i) of Example 3.5:

$$\sigma_{1,2} = \frac{3.205}{2} \pm [(\frac{3.205}{2})^2 + (2.155)^2]^{1/2}$$

or

$$\sigma_1 = 4.288 \text{ MPa}$$
$$\sigma_2 = -1.083 \text{ MPa}$$ ◄

(CONT.)

3.34 CONT.

(b) $\tau_{max} = \frac{1}{2}(4.288 + 1.083)$

$= 2.686$ MPa ◄

(c) $\sigma_{oct.} = \frac{1}{3}(4.288 - 1.083)$

$= 1.068$ MPa ◄

$$\tau_{oct.} = \frac{1}{3}[(4.288+1.083)^2 + (-1.083)^2 + (-4.288)^2]^{1/2}$$

$= 2.319$ MPa ◄

3.35

We apply Eqs. (3.58).

(a)

$$a = 0.88[\frac{2(500)(0.025\times0.0375)}{(200\times10^9)(0.0125)}]^{1/3}$$

$= 0.635$ mm ◄

(b)

$$\sigma_c = 0.62[500(200\times10^9)^2 \times (\frac{0.0125}{2\times0.025\times0.0375})^2]^{1/3}$$

$= 596.1$ MPa ◄

(c)

$$\delta = 1.54[(500)^2 \times \frac{0.0125}{2(200\times10^9)^2(0.025\times0.0375)}]^{1/3}$$

$5.339(10^{-3})$ mm ◄

3.36

(a) Use Eq. (3.57):

$$\sigma_c = 0.62[\frac{500(200\times10^9)^2}{4(0.025)^2}]^{1/3}$$

$= 1240$ MPa ◄

(b) Apply Eqs. (3.55) and (3.54) for $r_1 = r_2 = r$ and $E_1 = E_2 = E$ to obtain the formula

$$\sigma_c = 0.617[\frac{PE^2}{r^2}]^{1/3}$$

Thus,

$$\sigma_c = 0.617[\frac{500(200\times10^9)^2}{(0.025)^2}]^{1/3}$$

$= 1959$ MPa ◄

3.37

Using Eqs. (3.68), (3.69), and the second of (3.65), we have

$$m=\frac{4}{(1/0.4)+(1/0.25)}=0.6154$$

$$n=\frac{4(200\times10^9)}{3(1-0.3^2)}=2.9304(10^{11})$$

$$\cos\alpha=\pm\frac{(1/0.4)-(1/0.25)}{(1/0.4)+(1/0.25)}=0.2308$$

or

$$\alpha=76.66^\circ$$

Interpolating Table 3.3:

$c_a=1.1774 \qquad c_b=0.8616$

Apply Eqs. (3.64):

$$a=1.1774[\frac{4(10^3)(0.6154)}{2.9304\times10^{11}}]^{1/3}=2.393\text{ mm} \blacktriangleleft$$

$$b=0.8616[\frac{4(10^3)(0.6154)}{2.9304\times10^{11}}]^{1/3}=1.752\text{ mm} \blacktriangleleft$$

Thus,

$$\sigma_c=1.5\frac{4(10^3)}{\pi(2.393\times1.752)10^{-6}}=455.5\text{ MPa} \blacktriangleleft$$

3.38

Use Eqs. (3.62):

$$\sigma_c=0.418[\frac{2.5(10^3)(200\times10^9)}{0.1(0.005)}]^{1/2}=418\text{ MPa} \blacktriangleleft$$

$$2b=2\{1.52[\frac{2.5(10^3)(0.005)}{200\times10^9(0.1)}]^{1/2}\}$$

$$=2(0.038)=0.076\text{ mm} \blacktriangleleft$$

3.39

Equations (3.60), for $\nu_1=\nu_2=0.25$, $E_1=E_2=E$, $r_1=r_2=r$:

$$\sigma_c=0.412[\frac{2(10^6)200\times10^9(2)}{0.2}]^{1/2}=824\text{ MPa} \blacktriangleleft$$

$$2b=2\{1.545[\frac{2(10^6)(0.2)}{(200\times10^9)2}]^{1/2}$$

$$=2(1.545)=3.09\text{ mm} \blacktriangleleft$$

3.40

Refer to Example 3.6.

$1/r_1'=0 \qquad 1/r_2'=0 \qquad \theta=\pi/2$

$$m=\frac{4}{(1/0.5)+(1/0.2)}=0.5714$$

$$n=\frac{4(210\times10^9)}{3(1-0.25^2)}=2.9867(10^{11})$$

$$\cos\alpha=\pm\frac{(1/0.5)-(1/0.2)}{(1/0.5)+(1/0.2)}=0.4286$$

or

$\alpha=64.62^\circ$

Table 3.3: $c_a=1.3862$, $c_b=0.7758$

$$a=1.3862[\frac{5\times10^3(0.5714)}{2.9867\times10^{11}}]^{1/3}=2.942\text{ mm} \blacktriangleleft$$

$$b=0.7558[\frac{5\times10^3(0.5714)}{2.9867\times10^{11}}]^{1/3}=1.604\text{ mm} \blacktriangleleft$$

Thus,

$$\sigma_c=1.5\frac{5(10^3)}{\pi(2.942\times1.604)10^{-6}}=505.9\text{ MPa} \blacktriangleleft$$

3.41

Refer to Example 3.6. We now have $r_1=r_2=r$. Thus, Equations (3.68) and (3.69) become

$$m=\frac{4}{(1/r)+(1/r)}=2r=2(0.2)=0.4$$

$$\cos\alpha=\pm\frac{(1/r)-(1/r)}{(1/r)+(1/r)}=0,\quad \alpha=90^\circ$$

From Table 3.3 it can be concluded that surface of contact has a circular boundary: $c_a=c_b=1$.

Equation (3.65):

$$n=\frac{4(210\times10^9)}{3(1-0.25^2)}=2.98667(10^{11})$$

$$a=b=1[\frac{5(10^3)(0.4)}{2.98667\times10^{11}}]^{1/3}$$

$$=1.885\text{ mm} \blacktriangleleft$$

Thus

$$\sigma_c=1.5\frac{5(10^3)}{\pi(1.885)^2(10^{-6})}$$

$$=671.9\text{ MPa} \blacktriangleleft$$

3.42

Given quantities are:

$r_1 = r_1' = 0.025$ m $\qquad r_2 = -0.03$ m

$r_2' = -0.125$ m $\qquad \nu = 0.3$

E=200GPa

Thus,

$$m = \frac{4}{\frac{1}{0.025} + \frac{1}{0.025} - \frac{1}{0.3} - \frac{1}{0.125}} = 0.10345$$

$$n = \frac{4(200\times10^9)}{3(1-0.09)}$$

$$= 293.04029(10^9)$$

Also

$$A = \frac{2}{m} = \frac{2}{0.10345} = 19.33301$$

$$B = \frac{1}{2}[(0)^2 + (\frac{1}{r_2} - \frac{1}{r_2'}) + 2(0)]^{1/2}$$

$$= 12.66667$$

$$\alpha = \cos^{-1}\frac{12.66667}{19.33301} = 49.06645^\circ$$

From Table 3.3, we find

$c_a = 1.78611 \qquad c_b = 0.63409$

The semiaxes are then

$$a = 1.78611\left[\frac{1800(0.10345)}{293.04029(10^9)}\right]^{1/3}$$

$$= 0.00154 \text{ m} = 1.54 \text{ mm}$$

$$b = 0.63409\left[\frac{1800(0.10345)}{293.04029(10^9)}\right]^{1/3}$$

$$= 0.00055 \text{ m} = 0.55 \text{ mm}$$

Maximum contact pressure is therefore

$$\sigma_c = 1.5\frac{1800}{\pi(1.54\times0.55)(10^{-6})}$$

$$= 1014.7 \text{ MPa}$$ ◄

3.43

Now given data is as follows:

$r_1 = r_1' = 0.02$ m $\qquad r_2 = -0.022$ m

$r_2' = -0.125$ m $\qquad \nu = 0.3$

E=200 GPa

Therefore,

$$m = \frac{4}{\frac{1}{0.02} + \frac{1}{0.02} - \frac{1}{0.022} - \frac{1}{0.125}} = 0.08594$$

$$n = \frac{4(200\times10^9)}{3(1-0.09)}$$

$$= 293.0403(10^9)$$

We have

$$A = \frac{2}{m} = 23.2721$$

$$B = \pm\frac{1}{2}\left[-\frac{1}{0.022} + \frac{1}{0.125}\right]$$

$$= 18.7273$$

$$\alpha = \cos^{-1}\frac{18.7273}{23.2721} = 36.42^\circ$$

Using Table 3.3:

$c_a = 2.323 \qquad c_b = 0.541$

Then, the semiaxes are:

$$a = 2.323\left[\frac{1800(0.08594)}{293.0403\times10^9}\right]^{1/3}$$

$$= 0.00188 \text{ m} = 1.88 \text{ mm}$$

$$b = 0.541\left[\frac{1800(0.08594)}{293.0403\times10^9}\right]^{1/3}$$

$$= 0.00044 \text{ m} = 0.44 \text{ mm}$$

Maximum contact stress is now obtained as

$$\sigma_c = 1.5\frac{1800}{\pi(1.88\times0.44)10^{-6}}$$

$$= 1039 \text{ MPa}$$ ◄

4.1

State of stress is given by

$$\sigma_1 = \sigma = \frac{32M}{\pi(0.1)^3} + \frac{4P}{\pi(0.1)^2}, \qquad \sigma_2 = \sigma_3 = 0$$

Refer to Table 4.1 and Eq. (2.56):

$$0.47\sigma_{yp} = 0.47\sigma \quad \text{or} \quad \sigma_{yp} = \sigma$$

Thus,

$$221(10^3) = \frac{32(17)}{\pi(0.1)^3} + \frac{4P}{\pi(0.1)^2}$$

Solving,

$$P = 375.7 \text{ kN}$$ ◄

4.2

Referring to Appendix B, we obtain

$$\sigma_1 = 101.3 \text{ MPa} \qquad \sigma_2 = 0$$
$$\sigma_3 = -51.32 \text{ MPa}$$

(a) $$|\sigma_1 - \sigma_3| = \sigma_{yp}$$

or

$$\sigma_{yp} = |101.3+51.32| = 152.6 \text{ MPa}$$ ◄

(b)

$$\tau_{oct} = \frac{1}{3}[(101.3)^2 + (101.3+51.32)^2 + (51.32)^2]^{1/2} = 63.41 \text{ MPa}$$

Thus,

$$\sigma_{yp} = 63.41/0.47 = 134.9 \text{ MPa}$$ ◄

4.3

Using Eq. (4.6), we have

$$0.47\frac{\sigma_{yp}}{f_s} = \frac{1}{3}[(\sigma_1 - \sigma_2)^2 + \sigma_1^2 + \sigma_2^2]^{1/2} \quad \text{(a)}$$

Here

$$\sigma_{1,2} = \frac{\sigma}{2} \pm \frac{1}{2}[\sigma^2 + 4\tau^2]^{1/2}$$

and

$$\sigma = \frac{32M}{\pi d^3} + \frac{4P}{\pi d^2}, \qquad \tau = \frac{16T}{\pi d^3}$$

Substituting the given data:

$$\sigma_{1,2} = \frac{1}{2}\left[\frac{32(4\times10^3)}{\pi(0.12)^3} + \frac{4(45\times10^3)}{\pi(0.12)^2}\right] \pm$$
$$\frac{1}{2}\left\{\left[\frac{32(4\times10^3)}{\pi(0.12)^3} + \frac{4(45\times10^3)}{\pi(0.12)^2}\right]^2 + 4[16(11.2\times10^3)/\pi(0.12)^3]^2\right\}^{1/2}$$

or

$$\sigma_1 = 49.55 \text{ MPa} \qquad \sigma_2 = -21.99 \text{ MPa}$$

Equation (a) is then

$$1.41\frac{280}{f_s} = [(71.54)^2 + (49.55)^2 + (-21.99)^2]^{1/2}$$

Solving,

$$f_s = 4.4$$ ◄

4.4

Maximum stresses, occurring at the fixed end are:

$$\sigma = \frac{Mc}{I} = \frac{450(0.25t)}{2t^4/3} = \frac{168.75}{t^3}$$

$$\tau = \frac{3}{2}\frac{450}{2t^2} = \frac{337.5}{t^2}$$

From Eq. (4.9a), we have

$$\sigma_{yp}^2 = \sigma^2 + 3\tau^2 = (280\times10^6)^2$$

Therefore, at neutral axis $\sigma = 0$:

$$\sigma_{yp} = \sqrt{3}\,\tau \quad \text{gives} \quad t = 1.45 \text{ mm}$$

At the extreme fibers, $\tau = 0$:

$$\sigma_{yp} = \sigma \quad \text{gives} \quad t = 8.45 \text{ mm}$$

Allowable width is thus

$$t_{all} = 8.45 \text{ mm}$$ ◄

4.5

We have

$$\tau_{yp} = \frac{\sigma_{yp}}{2} = 175 \text{ MPa}, \quad \sigma_1 = -\sigma_2 = \tau$$

(a)

$$\frac{175}{1.5} = \frac{16(500)}{\pi d^3}$$

or

$$d = 27.95 \text{ mm}$$ ◄

(b)

$$\sigma_{1,2} = \frac{\sigma}{2} \pm \frac{1}{2}\sqrt{\sigma^2 + 4\tau^2}$$

Here

$$\sigma = \frac{32M}{\pi d^3} = 32(pL^2/8)/\pi d^3$$
$$= \frac{32}{8\pi d^3}\left[\frac{1}{4}\pi d^2(77\times10^3)10^2\right]$$
$$= 77(10^5)/d$$
$$\tau = \frac{16T}{\pi d^3} = \frac{16(500)}{\pi d^3} = 2547.77/d^3$$

Using Eq. (4.8a):

$$\frac{\sigma_{yp}}{f_s} = \sqrt{\sigma^2 + 4\tau^2}$$

$$\frac{350(10^6)}{1.5} = \left[\left(\frac{77\times10^5}{d}\right)^2 + 4\left(\frac{2547.77}{d^3}\right)^2\right]^{1/2}$$

or

$$5.444(10^{16}) = \frac{5.929(10^{13})}{d^2} + \frac{2.594(10^7)}{d^6}$$

Solving, by trial and error:

$$d = 0.0368 \text{ m} = 36.8 \text{ mm}$$ ◄

4.6

We have σ_{all} =90/1.2=75 MPa

(a)

$$\sigma_{all} = |\sigma_1 - \sigma_3|$$

$$=63.4+12.2$$

$$=75.6>75 \quad \text{Failure occurs} \blacktriangleleft$$

(b)

$$2\sigma_{all}^2=(63.4-0.53)^2+(0.53+12.2)^2 +(-12.2-63.4)^2$$

or

$$\sigma_{all}=70<75 \quad \text{No failure} \blacktriangleleft$$

4.7

(a) Using the torsion formula,

$$\tau = \frac{T(0.05)}{\pi(0.05)^4/2} = 5093T$$

and

$$\sigma_1 = -\sigma_2 = 5093T$$

Equation (4.5a) yields then

$$[280(10^6)]^2=(5093T)^2+ (5093T)(5093T)+(-5093T)^2$$

Solving,

$$T=31.74 \text{ kN·m} \blacktriangleleft$$

(b) We now have

$$\sigma = \frac{400(10^3)\pi}{\pi(0.05)^2} = 160 \text{ MPa}$$

$$=5093T \quad \text{(as before)}$$

Principal stresses are:

$$\sigma_{1,2} = \underbrace{\frac{160(10^6)}{2}}_{a} \pm \underbrace{\frac{1}{2}[(160\times10^6)^2+4(5093T)^2]^{1/2}}_{b}$$

$$=a \pm b$$

With this notation, Eq. (4.5a) becomes

$$\sigma_{yp}^2=(a+b)^2-(a^2-b^2)+(a-b)^2=a^2+3b^2$$

Thus,

$$(280\times10^6)^2=\left(\frac{160\times10^6}{2}\right)^2+ \frac{3}{4}[(160\times10^6)^2+4(5093T)^2]$$

Solving,

$$T=26.05 \text{ kN·m} \blacktriangleleft$$

4.8

Maximum moment is

$$M=\frac{PL^2}{8}=\frac{6(1.5)^2}{8}=1.688 \text{ kN·m}$$

and hence

$$\sigma = \sigma_1 = \frac{My}{I} = \frac{1.688(0.125)10^3}{0.1(0.25)^3/12}$$

$$=1.62 \text{ MPa}$$

(a)

$$\frac{\sigma_{yp}}{f_s} = \sqrt{\sigma_1^2 - 0} = \sigma_1$$

$$\frac{28}{f_s}=1.62$$

from which

$$f_s=17.3 \blacktriangleleft$$

(b)

$$|\sigma_1 - 0| = \frac{\sigma_{yp}}{f_s}$$

or

$$f_s=17.3 \blacktriangleleft$$

4.9

Referrring to Appendix B, we compute

$$\sigma_1 = 12.05 \text{ MPa}$$
$$\sigma_2 = -1.521 \text{ MPa}$$
$$\sigma_3 = -4.528 \text{ MPa}$$

Using Eq. (4.5a),

$$2\sigma_{yp}^2=(12.05+1.521)^2+(-1.521+4.528)^2 +(-4.528-12.05)^2$$

$$=468$$

Solving,

$$\sigma_{yp}=15.3 \text{ MPa}$$

Hence,

$$\tau_{yp}=15.3(0.577)=8.828 \text{ MPa}$$

Therefore

$$\sigma_y=2(9/8.828)=2.039 \text{ MPa}$$

$$\sigma_x=3(9/8.828)=3.058 \text{ MPa} \blacktriangleleft$$

4.10

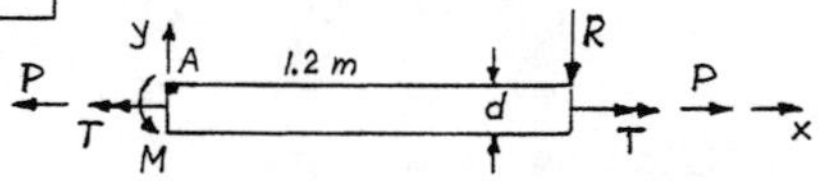

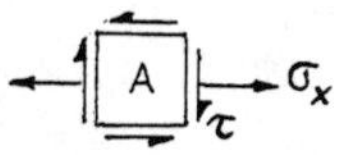

We have P=50R, T=0.8R, and M=1.2R. Stresses are

$$\sigma_b=\frac{32M}{\pi d^3}=\frac{32\,(1.2R)}{\pi(0.05)^3}=97{,}784.8R$$

$$\tau=-\frac{16T}{\pi d^3}=\frac{-16\,(0.8R)}{\pi\,(0.05)^3}=-32{,}595R$$

$$\sigma_a=\frac{50R}{\pi\,(0.05)^2/4}=25{,}464.8R$$

$$\sigma_x=\sigma_a+\sigma_b=123{,}249.6R$$

(a) Equation (4.8a):

$$260(10^6)/2=R[(123{,}249.6)^2+4(-32{,}595)^2]^{1/2}$$

or

$$R=932\text{ N}$$

(b) Equation (4.9a):

$$130(10^6)=R[(123{,}249.6)^2+3(-32{,}595)^2]^{1/2}$$

or

$$R=959\text{ N}$$

4.11

Referring to Appendix B,

σ_1 =197.4 MPa σ_2 =-14.44 MPa σ_3 =-72.96 MPa

Applying Eq. (4.4b):

$$2\sigma_{yp}^2=(197.4+14.44)^2+(-14.44+72.96)^2+(-72.96-197.4)^2$$

or

$$\sigma_{yp}=246.4\text{ MPa}$$

Hence,

$$\tau_{yp}=246.4(0.577)=142.2\text{ MPa}$$

Thus,

$$\sigma_y=40\,\frac{140}{142.2}=39.38\text{ MPa}$$

$$\sigma_x=50\,\frac{140}{142.2}=49.23\text{ MPa}$$

4.12

Stresses are

σ_1 =pr/t=p(0.25)/0.005=50p

σ_2 =pr/2t=25p

(a) Applying Eq. (4.5a),

$$(50p)^2-1250p^2+(25p)^2=(280)^2$$

or

$$p=6.466\text{ MPa}$$

(b) Using Eq. (4.2a),

$$|50p-0|=280$$

or

$$p=5.6\text{ MPa}$$

4.13

We have $(\sigma_{yp})_{all}$ =82/1.2=68.33 MPa

Referrring to Appendix B:

σ_1 =63.44 MPa σ_2 =0.533 MPa σ_3 =-12.17 MPa

(a) Using Eq. (4.1),

$$\sigma_{yp}=|63.44+12.17|=75.6>68.33:\ \text{Failure occurs}$$

(b) Applying Eq. (4.4b),

$$2\sigma_{yp}^2=(63.44-0.533)^2+(0.533+12.17)^2+(-12.17-63.44)^2$$

or

$$\sigma_{yp}=70.13>68.33:\ \text{Failure occurs}$$

4.14

Referring to Appendix B:

σ_1 =162.4 MPa σ_2 =46.15 MPa σ_3 =1.468 MPa

(a) Equation (4.1) yields

$$f_s=\frac{300}{162.4-1.468}=1.86$$

(b) Equation (4.4b) gives

$$f_s^2=\frac{2\sigma_{yp}^2}{(\sigma_1-\sigma_2)^2+(\sigma_2-\sigma_3)^2+(\sigma_3-\sigma_1)^2}=\frac{2(300)^2}{(116.25)^2+(44.682)^2+(160.932)^2}$$

or

$$f_s=2.09$$

4.15

Referring to Appendix B:

$\sigma_1=156.2$ MPa $\quad \sigma_2=42.13$ MPa

$\sigma_3=11.7$ MPa

(a) Equation (4.1) gives

$$f_s=\frac{220}{156.2-11.7}$$

$$=1.52 \quad ◄$$

(b) Equation (4.4b) yields

$$f_s^2=\frac{2\sigma_{yp}^2}{(\sigma_1-\sigma_2)^2+(\sigma_2-\sigma_3)^2+(\sigma_3-\sigma_1)^2}$$

$$=\frac{2(220)^2}{(114.07)^2+(30.43)^2+(-144.5)^2}$$

or

$$f_s=1.67 \quad ◄$$

4.16

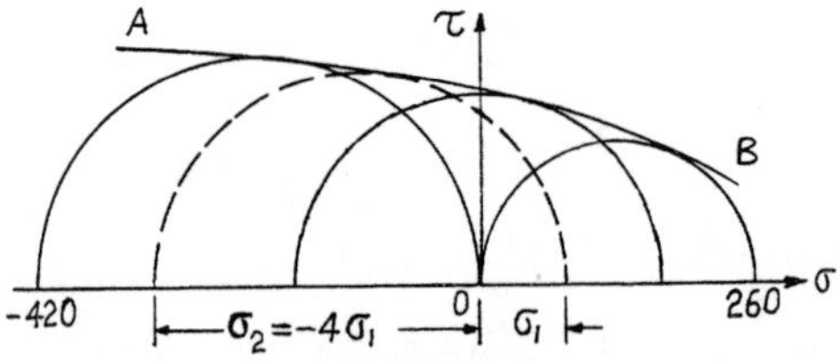

(a) Upon following the procedure described in Sec. 4.10, Mohr's circle is constructed as shown in the sketch above.

The circle representing the given loading is then drawn by a trial and error procedure, as is indicated by the dashed lines. From the diagram, we measure the following values:

$\sigma_1=77$ MPa $\quad \sigma_2=-308$ MPa ◄

(b)

Applying Eq. (4.12a),

$$\frac{\sigma_1}{\sigma_u}-\frac{\sigma_2}{\sigma_u'}=1$$

or

$$\frac{\sigma_1}{260}-\frac{-4\sigma_1}{420}=1$$

Solving,

$\sigma_1=75$ MPa $\quad \sigma_2=-300$ MPa ◄

4.17

Principal stresses are

$$\sigma_{1,2}=\frac{-180}{2}\pm[(\frac{180}{2})^2+200^2]^{1/2}$$

or

$\sigma_1=129.3$ MPa, $\sigma_2=-309.3$ MPa

(a) Equations (4.11a):

$|\sigma_1|<290$ MPa

But since

$|\sigma_2|>290$ MPa: failure occurs ◄

(b) Equation (4.12a):

$$\frac{129.3}{290}-\frac{-309.3}{650}=1$$

gives $0.446+0.476=0.922<1$

Thus, no fracture ◄

Note that Coulomb-Mohr theory is the most reliable when $\sigma_u' \gg \sigma_u$, as in this example.

4.18

$$\sigma_x=\frac{pr}{2t}+\frac{P}{2\pi rt}$$

$$=\frac{2.8(10^6)125}{2(5)}+\frac{45(10^3)}{2\pi(0.125)(0.005)}$$

$=46.46$ MPa

$$\sigma_y=\frac{pr}{t}=\frac{2.8(10^6)125}{5}=70 \text{ MPa}$$

$$\tau=\frac{Tr}{2\pi r^3 t}=\frac{31.36(10^3)}{2\pi(0.125^2)(0.005)}=63.89 \text{ MPa}$$

Thus,

$$\sigma_{1,2}=\tfrac{1}{2}(46.46+70)\pm[\tfrac{1}{4}(46.46-70)^2+(63.89)^2]^{1/2}$$

or

$\sigma_1=123.2$ MPa $\quad \sigma_2=-6.74$ MPa

(a) Equation (4.12a),

$$\frac{123.2}{210}-\frac{-6.74}{500}=1$$

gives $0.587+0.013=0.6<1$

Thus, no fracture ◄

(b) Equations (4.11a) shows

$123.2<210$, no fracture

$6.14<210$, no fracture ◄

4.19

State of stress is represented by Mohr's circle shown below.

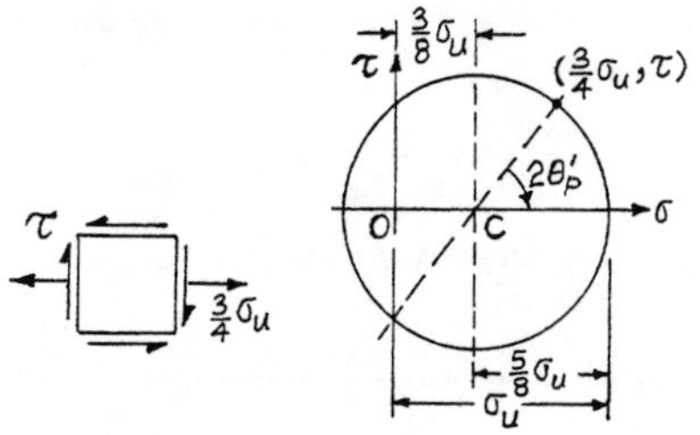

From the circle, we obtain

$$\tau = \sigma_u \sqrt{(5/8)^2 - (3/8)^2} = \frac{1}{2}\sigma_u$$

and

$$\theta_p' = \frac{1}{2}\tan^{-1}\frac{1/2}{3/8} = 26.57° \quad ◄$$

Orientation of the fracture plane is shown below.

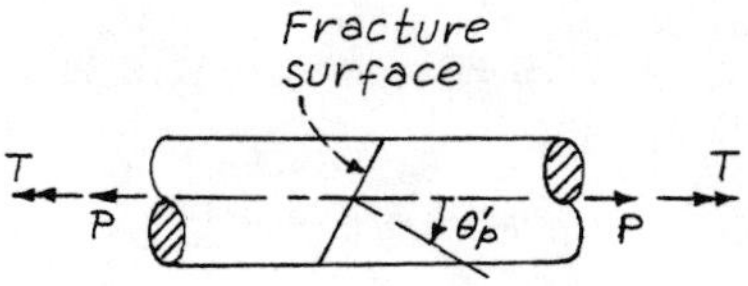

4.20

Principal stresses are

$$\sigma_{1,2} = \frac{1}{2}(200+20) \pm [(90)^2 + (150)^2]^{1/2}$$

or

$$\sigma_1 = 284.9 \text{ MPa} \qquad \sigma_2 = -64.9 \text{ MPa}$$

(a)

$$284.9 = 420/f_s, \qquad f_s = 1.47 \quad ◄$$

$$64.9 = 420/f_s, \qquad f_s = 6.47$$

(b) Equation (4.12a):

$$\frac{284.9}{420} - \frac{-64.9}{900} = \frac{1}{f_s}$$

Solving,

$$f_s = 1.33 \quad ◄$$

4.21

Uniform shear stress τ acts on a typical element as shown.

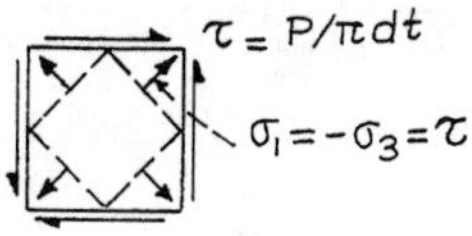

(a)

$$|\sigma_1| = \sigma_u \quad \text{or} \quad |\sigma_3| = \sigma_u$$

$$\frac{P}{\pi t d} = \sigma_u, \qquad P = \pi t d \sigma_u \quad ◄$$

(b)

$$\frac{\sigma_1}{\sigma_u} - \frac{\sigma_3}{\sigma_u'} = 1; \qquad \tau\left(1 + \frac{\sigma_u}{\sigma_u'}\right) = \sigma_u$$

or

$$P = td\sigma_u / (1 + \sigma_u/\sigma_u') \quad ◄$$

4.22

$$\sigma_m = 120(10^3)/25(10^{-4}) = 48 \text{ MPa}$$

and

$$\frac{\sigma_a}{\sigma_{cr}} + \frac{\sigma_m}{\sigma_f} = 1; \qquad \frac{1200 F_A}{240(10^6)} + \frac{48(10^6)}{700(10^6)} = 1$$

or

$$F_A = 186.3 \text{ kN} \quad ◄$$

4.23

$$\sigma_{1a} - \sigma_{3a} = 15p - 0 = \sigma_{ea}$$
$$\sigma_{1m} - \sigma_{3m} = 9p - 0 = \sigma_{em}$$

Then,

$$\frac{15p}{250(10^6)} + \frac{9p}{300(10^6)} = 1$$

or

$$p = 11.11 \text{ MPa} \quad ◄$$

4.24

We have

$$\sigma_{cr} = \frac{\sigma_a}{1 - (\sigma_m/\sigma_u)} \qquad \text{(a)}$$

where,

$$\sigma_m = \frac{F_{max} + F_{min}}{2A}, \quad \sigma_a = \frac{F_{max} - F_{min}}{2A} \qquad \text{(b)}$$

Substituting Eqs. (b) into (a):

$$\sigma_{cr} = \frac{(F_{max} - F_{min})/2A}{1 - [(F_{max} + F_{min})/2A\sigma_u]}$$

Solving,

$$A = \frac{1}{\sigma_{cr}}\left[F_{max} - \frac{1}{2}(F_{max} + F_{min})\left(1 - \frac{\sigma_{cr}}{\sigma_u}\right)\right] \quad ◄$$

4.25

We have $\sigma_m = \sigma_a$. Using Table 4.2:

$$\frac{\sigma_m}{510/1.5} + \frac{\sigma_m}{1050/1.5} = 1$$

from which σ_m=228.8 MPa.

At the fixed end:

$$M_{max} = PL = 10(0.05) = 0.5 \text{ N·m}$$

Hence,

$$M_{a,m} = (M_{max} \pm M_{min})/2 = 0.25 \text{ N·m}$$

and

$$\sigma_a = \sigma_m = \frac{6M_m}{bt^2} = \frac{6(0.25)}{0.005t^2} = \frac{300}{t^2} = 228.8(10^6)$$

Solving,

t=1.145 mm ◄

4.26

We have $\sigma_a = \sigma_m$. From Table 4.2:

$$\frac{\sigma_m}{740/2.5} + \frac{\sigma_m}{1500/2.5} = 1$$

or σ_m=198.2 MPa.

At the center of the beam:

$$M_{max} = PL/4 = 0.25(20)(0.125) = 0.625 \text{ N·m}$$

Hence,

$$M_{a,m} = (M_{max} \pm M_{min})/2 = 0.3125 \text{ N·m}$$

and

$$\sigma_a = \sigma_m = \frac{6M_m}{bt^2} = \frac{6(0.3125)}{0.01t^2} = \frac{187.5}{t^2} = 198.2(10^6)$$

from which

t=0.973 mm ◄

4.27

$$\sigma_{max} = \sigma_{min} = \frac{Mc}{I}, \quad \sigma_m = \frac{4M}{\pi r^3}, \quad \sigma_a = 0$$

$$\tau_{max} = \frac{Tr}{J}, \quad \tau_{min} = 0, \quad \tau_m = \tau_a = \frac{Tr}{\pi r^3}$$

and

$$\sigma_x = \sigma, \quad \tau_{xy} = \tau, \quad \sigma_y = \sigma_z = \tau_{xz} = \tau_{yz} = 0$$

Equations (4.16) yield

$$\sqrt{0+3\tau_a^2} = \sigma_{ea}, \quad \sqrt{\sigma_m^2 + 3\tau_m^2} = \sigma_{em}$$

Then, Soderberg relation becomes

(CONT.)

4.27 CONT.

$$\sigma_{cr} = \frac{\sqrt{3}\,\tau_a}{1 - \frac{\sqrt{\sigma_m^2 + 3\tau_m^2}}{\sigma_{yp}}}$$

or

$$\sigma_{cr} = \frac{\sigma_{cr}}{\sigma_{yp}}\sqrt{\sigma_m^2 + 3\tau_m^2} + \sqrt{3}\,\tau_a$$

or

$$\sigma_{cr} = \frac{\sigma_{cr}}{\sigma_{yp}}\left[\left(\frac{4M}{\pi r^3}\right)^2 + 3\left(\frac{T}{\pi r^3}\right)^2\right]^{1/2} + \frac{\sqrt{3}T}{\pi r^3}$$

Solving this expression for r, we obtain Eq. (P4.27).

4.28

$$\sigma_{1a,2a} = \frac{\sigma_{xa} + \sigma_{ya}}{2} \pm \left[\left(\frac{\sigma_{xa} - \sigma_{ya}}{2}\right)^2 + \tau_{xya}^2\right]^{1/2}$$

$$= \frac{\sigma_{xa}}{2} \pm \left[\frac{\sigma_{xa}^2}{4} + \tau_{xya}^2\right]$$

$$= \frac{900}{2} \pm \left[\frac{81(10^4)}{4} + 4(10^4)\right]^{1/2}$$

or

σ_{1a} =942.44 MPa, σ_{2a}=-42.44 MPa

Similarly,

σ_{1m} =161.8 MPa, σ_{2m}=-61.8 MPa

Thus,

$$\sigma_{ea} = \sigma_{1a} - \sigma_{2a} = 984.88 \text{ MPa}$$
$$\sigma_{em} = \sigma_{1m} + \sigma_{2m} = 223.6 \text{ MPa}$$

(a) Modified Goodman relation:

$$\sigma_{cr} = \frac{984.88}{1-(223.6/2400)} = 1086 \text{ MPa}$$

$$b = \frac{\ln(0.9 \times 2400/800)}{\ln(10^3/10^8)} = -0.0863$$

$$N_{cr} = 10^3(1086/0.9\times2400)^{-11.587} = 2.89(10^6) \text{ cycles}$$ ◄

(b) Soderberg criterion:

$$\sigma_{cr} = \frac{948.88}{1-(223.6/1600)} = 1145 \text{ MPa}$$

$$N_{cr} = 10^3(1145/0.9\times2400)^{-11.587} = 1.57(10^6) \text{ cycles}$$ ◄

(c) The SAE criterion:

σ_{cr}=1086 MPa, b=-0.0596

$$N_{cr} = 1(1086/2400)^{-16.778} = 0.59(10^6) \text{ cycles}$$ ◄

(d) Gerber criterion:

$$\sigma_{cr} = \frac{948.88}{1-(223.6/2400)^2} = 993.51 \text{ MPa}$$

$$N_{cr} = 10^3(993.51/0.9\times2400)^{-11.587} = 8.12(10^6) \text{ cycles}$$ ◄

4.29

$\sigma_{xa}=(800+600)/2=700$ MPa
$\sigma_{xm}=(800-600)/2=100$ MPa
$\sigma_{ya}=(500+300)/2=400$ MPa
$\sigma_{ym}=(500-300)/2=100$ MPa
$\tau_{xya}=(200+150)/2=175$ MPa
$\tau_{xym}=(200-150)/2=25$ MPa

Equations (4.16a) give then

$$2\sigma_{ea}^2=(700-400)^2+400^2+700^2+6(175)^2$$
$$2\sigma_{em}^2=(200-100)^2+100^2+100^2+6(25)^2$$

or

$\sigma_{ea}=679.61$ MPa, $\sigma_{em}=108.97$ MPa

(a) Modified Goodman criterion:

$$\sigma_{cr}=\frac{679.61}{1-(108.97/1600)}=729.27 \text{ MPa}$$
$$b=\frac{\ln(0.9\times1600/0.5\times1600)}{\ln(10^3/10^3)^6}=-0.08509$$
$$N_{cr}=10^3(729.27/0.9\times1600)^{-11.752}$$
$$=2.97(10^6) \text{ cycles}$$ ◄

(b) Soderberg criterion:

$$=\frac{679.61}{1-(108.97/1000)}=762.72 \text{ MPa}$$
$$N_{cr}=10^3(762.72/0.9\times1600)^{-11.752}$$
$$=1.75(10^6) \text{ cycles}$$ ◄

4.30

$$\sigma_a=(M_{max}-M_{min})c/2I$$
$$\sigma_m=(M_{max}+M_{min})c/2I, \qquad M=PL$$

Substituting these into Soderberg relation, we obtain

$$M_{max}=\frac{2\sigma_{cr}I}{(1+\lambda)c}+\frac{1-\lambda}{1+\lambda}M_{min}$$

where

$$\lambda=\sigma_{cr}/\sigma_{yp}=2/3$$

Thus,

$$P_{max}=\frac{2\sigma_{cr}I}{c(1+\lambda)L}+\frac{1-\lambda}{1+\lambda}P_{min}$$
$$=\frac{2(200\times10^6)(0.05\times0.1^3)}{12(5/3)(0.05)1.2}+\frac{1/3}{5/3}(10,000)$$
$$=16,667+2000=18.7 \text{ kN}$$ ◄

4.31

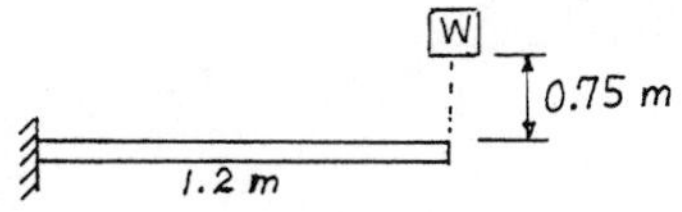

We have $\sigma_{yp}=Mc/I=PLc/I$
from which

$$P=\frac{\sigma_{yp}I}{Lc}=\frac{280(10^6)(0.05^4/12)}{1.2(0.025)}=4.86 \text{ kN}$$

End deflection is

$$\delta_{max}=\frac{4860(1.2)^3(12)}{3(200\times10^9)(0.05)^4}=26.86 \text{ mm}$$

But

$$W(h+\delta_{max})=\frac{1}{2}P\delta_{max}$$
$$W(0.75+0.02686)=\frac{1}{2}(4860)(0.02686)$$

Solving,

$$W=84.02 \text{ N}$$ ◄

4.32

The kinetic energy is

$$E_k=W\omega^2r^2/2g$$
$$=\frac{1090(240\times2\pi/60)^2(0.35)^2}{2(9.81)}=4300 \text{ N·m}$$

But

$$E_k=\frac{1}{2}T\theta \qquad \text{where } T=GJ\theta/L$$

Thus,

$$\theta=\sqrt{2LE_k/JG} \tag{a}$$

Substituting the data given:

$$\theta=\left[\frac{2\times1.5(4300)2}{80.5(10^9)\pi(0.0625)^4}\right]^{1/2}$$ ◄
$$=0.08309 \text{ rad}=4.76^\circ$$

Introducing

$$\theta=\frac{TL}{GJ}=\frac{\tau J}{r}\frac{L}{GJ}=\frac{\tau L}{rG}$$

into Eq. (a), we obtain

$$\tau=\sqrt{4GE_k/AL} \qquad \text{where } A=\pi c^2$$

Substituting the numerical values,

$$\tau=\left[\frac{4(80.5\times10^9)4300}{\pi(0.0625)^4(1.5)}\right]^{1/2}$$
$$=274.3 \text{ MPa}$$ ◄

5.1

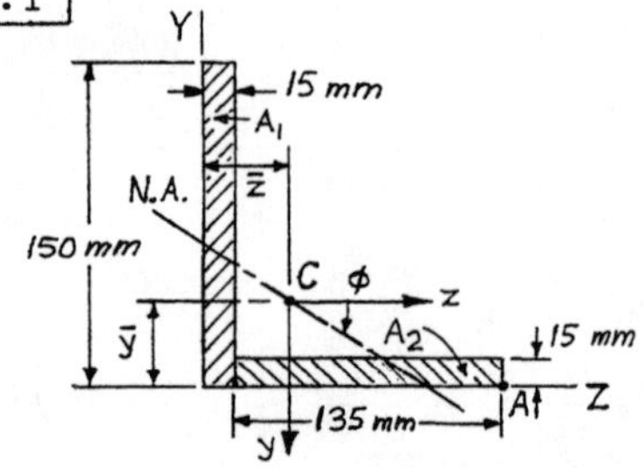

$\bar{z}=(A_1 z_1+A_2 z_2)/(A_1+A_2)$

$=\dfrac{(150\times15)7.5+(135\times15)[15+(135/2)]}{150\times15+135\times15}$

or

$\bar{z}=\bar{y}=43$ mm

Then,

$I_y=\frac{1}{12}(150)(15)^3+(150\times15)(35.5)^2$

$+\frac{1}{12}(15)(135)^3$

$+(135\times15)(39.5)^2$

or

$I_y=I_z=9.11(10^6)$ mm^4

$I_{yz}=(150\times15)(-32)(-35.5)$

$+(135\times15)(35.5)(39.5)$

$=5.4(10^6)$ mm^4

We have

$M_y=0,\ M_z=-11.25(0.9)=-10.125$ kN·m

Thus,

$(\sigma_x)_A=\dfrac{-10125(5.4)(0.107)+10125(9.11)(0.043)}{[(9.11)^2-(5.4)^2]\,10^{-6}}$

$=-35$ MPa ◄

Equation (5.15) gives

$\tan\phi=5.4/9.11=0.593$

or

$\phi=30.66°$ ◄

5.2

$I_y=2[\dfrac{30\times8^3}{12}+240\times34^2]+\dfrac{8\times76^3}{12}$

$=85(10^4)$ mm^4

$I_z=2[\dfrac{8\times30^3}{12}+240\times19^2]+\dfrac{76\times8^3}{12}$

$=21.2(10^4)$ mm^4

$I_{yz}=[0+240(19)(-34)]+$
$[0+240(-19)(34)]+0$
$=-31(10^4)$ mm^4

(CONT.)

5.2 CONT.

Using Eq. (5.14):

$M_o[21.2+1.5(-31)]z=M_o[-31+1.5\times85]y$

or $z=-3.81y$

Thus, point A is the farthest from the N.A., as shown.

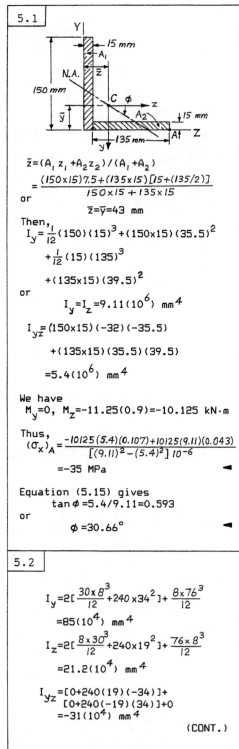

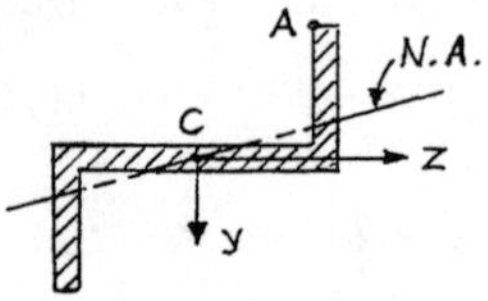

Hence, Eq. (5.13) gives

$(\sigma_x)_A=80(10^6)=$

$\dfrac{[21.2+1.5(-31)]0.03-[-31+1.5(85)](-0.034)}{[85\times21.2-(-31)^2]10^{-8}}M_o$

from which

$M_o=266.8$ N·m ◄

5.3

Bending moment at the midspan is

$M_z=-24(4)-24(2)=-48$ kN·m

From Fig. 5.4 and Example 5.1:

$z_E=0.105$ m $\qquad z_D=0$
$y_E=-0.045$ m $\qquad y_D=-0.045$ m

$I_y=I_z=11.596(10^{-6})$ m^4

$I_{yz}=-6.79(10^{-6})$ m^4

Then, Eq. (5.13) with M =0:

$(\sigma_x)_D=\dfrac{0-(-48000)(11.596)(-0.045)}{[(11.596)^2-(6.79)^2]10^{-6}}$

$=-283.4$ MPa ◄

Similarly,

$(\sigma_x)_E=\dfrac{-48000[-6.79(0.105)-11.596(-0.045)]}{[(11.596)^2-(6.79)^2]10^{-6}}$

$=103.8$ MPa ◄

5.4

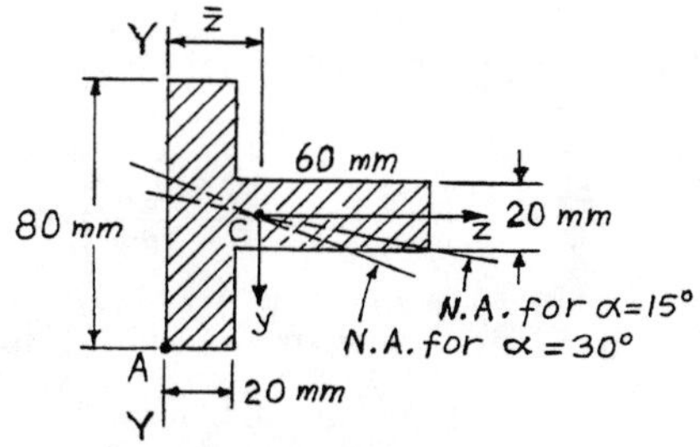

$$\bar{z} = \frac{80\times20(10)+60\times20(50)}{80\times20+60\times20}$$
$=27.14$ mm

$$I_y = \frac{80\times20^3}{12}+20\times80(17.14)^2+\frac{20\times60^3}{12}+20\times60(22.86)^2$$
$=15.1048(10^5)$ mm^4

$$I_z = \frac{80\times20^3}{12}+2\left[\frac{20\times30^3}{12}+20\times30(25)^2\right]$$
$=8.933(10^5)$ mm^4

Due to the symmetry $I_{yz}=0$.

(a) We have
$\alpha=0°$ $\quad M_y=0 \quad M_z=1.5P$

Equation (5.13) is thus,

$$\frac{290(10^6)}{1.2}=\frac{1.5Py}{I_z}=\frac{1.5P(0.04)}{8.933(10^{-7})}$$
or
$$P=3.6 \text{ kN}$$ ◄

(b) Now we have $\alpha=15°$ and
$M_z=1.5P\cos15°=1.4489P$
$M_y=1.5P\sin15°=0.3882P$

Equation (5.14):
$0.3882P(8.933)z$
$=1.4489P(15.1048)y$
from which
$z=6.3105y$

The farthest point from the N.A. is A. Equation (5.13):

$$(\sigma_x)_A=290(10^6)/1.2=\frac{0.3882P(8.933)(-0.02714)-1.4889P(15.1048)0.04}{8.933(15.1048)10^{-7}}$$

Solving,
$$P=3.36 \text{ kN}$$ ◄

5.5

Given $\alpha=30°$ and

$M_z=1.5P\cos30°=1.299P$
$M_y=1.5P\sin30°=0.75P$

The area properties are already found in Solution of Prob. 5.4. Equation (5.14) gives

$0.75P(8.933)z=$
$1.299P(15.1048)y$

or
$z=2.9287y$

As before, the maximum stress occurs at A. Equation (5.13):

$$(\sigma_x)_A=290(10^6)/1.2=\frac{0.75P(8.933)(-0.02714)-1.299P(15.1048)0.04}{8.933(15.1048)10^{-7}}$$

Solving,
$$P=3.37 \text{ kN}$$ ◄

5.6

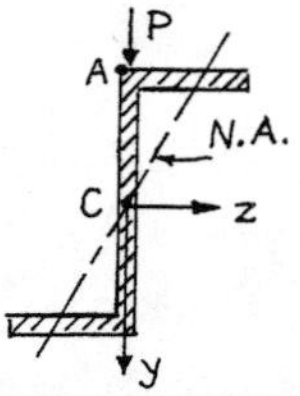

We have $\quad M_y=0 \quad$ and $\quad M_z=PL$

Equation (5.14) becomes

$$I_{yz}z=I_y y \quad \text{or} \quad -th^3z=\frac{2}{3}th^3y$$
or
$$y=-3z/2$$

Point A is the farthest from the N.A. Thus, with

$$y_A=-h-t/2 \quad \text{and} \quad z_A=-t/2$$

Equation (5.13) yields

$$(\sigma_x)_A=\frac{PL[-th^3(-t/2)-(2th^3/3)h^3(-h-t/2)}{(2th^3/3)(8th^3/3)-(-th^3)^2}$$

or
$$(\sigma_x)_A=\frac{3PL(2.5t+2h)}{7th^3}=\sigma_{max}$$ ◄

5.7

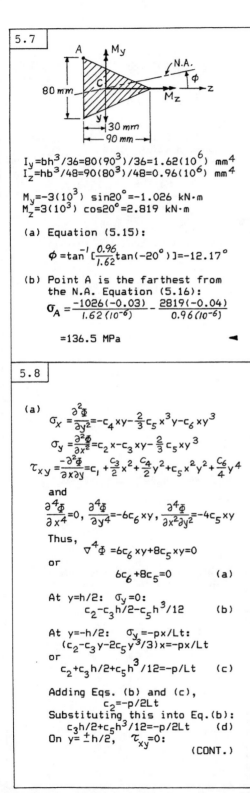

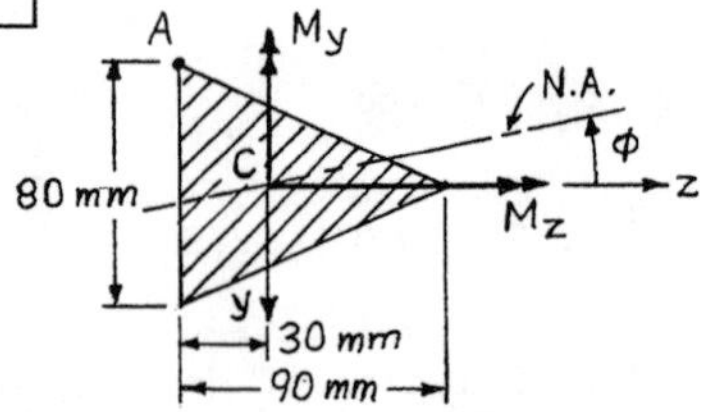

$I_y = bh^3/36 = 80(90^3)/36 = 1.62(10^6)$ mm^4
$I_z = hb^3/48 = 90(80^3)/48 = 0.96(10^6)$ mm^4

$M_y = -3(10^3)\sin 20^\circ = -1.026$ kN·m
$M_z = 3(10^3)\cos 20^\circ = 2.819$ kN·m

(a) Equation (5.15):

$$\phi = \tan^{-1}\left[\frac{0.96}{1.62}\tan(-20^\circ)\right] = -12.17^\circ$$

(b) Point A is the farthest from the N.A. Equation (5.16):

$$\sigma_A = \frac{-1026(-0.03)}{1.62(10^{-6})} - \frac{2819(-0.04)}{0.96(10^{-6})}$$

$= 136.5$ MPa ◄

5.8

(a)

$$\sigma_x = \frac{\partial^2\Phi}{\partial y^2} = -c_4 xy - \frac{2}{3}c_5 x^3 y - c_6 xy^3$$

$$\sigma_y = \frac{\partial^2\Phi}{\partial x^2} = c_2 x - c_3 xy - \frac{2}{3}c_5 xy^3$$

$$\tau_{xy} = \frac{-\partial^2\Phi}{\partial x\partial y} = c_1 + \frac{c_3}{2}x^2 + \frac{c_4}{2}y^2 + c_5 x^2 y^2 + \frac{c_6}{4}y^4$$

and

$$\frac{\partial^4\Phi}{\partial x^4} = 0,\quad \frac{\partial^4\Phi}{\partial y^4} = -6c_6 xy,\quad \frac{\partial^4\Phi}{\partial x^2\partial y^2} = -4c_5 xy$$

Thus,

$$\nabla^4\Phi = 6c_6 xy + 8c_5 xy = 0$$

or

$$6c_6 + 8c_5 = 0 \qquad \text{(a)}$$

At $y = h/2$: $\sigma_y = 0$:

$$c_2 - c_3 h/2 - c_5 h^3/12 \qquad \text{(b)}$$

At $y = -h/2$: $\sigma_y = -px/Lt$:

$$(c_2 - c_3 y - 2c_5 y^3/3)x = -px/Lt$$

or

$$c_2 + c_3 h/2 + c_5 h^3/12 = -p/Lt \qquad \text{(c)}$$

Adding Eqs. (b) and (c),

$$c_2 = -p/2Lt$$

Substituting this into Eq.(b):

$$c_3 h/2 + c_5 h^3/12 = -p/2Lt \qquad \text{(d)}$$

On $y = \pm h/2$, $\tau_{xy} = 0$:

(CONT.)

5.8 CONT.

$$-c_1 - \frac{c_3}{2}x^2 - \frac{c_4}{8}h^2 - \frac{c_5}{4}h^2 x^2 - \frac{c_6}{64}h^4 = 0$$

or

$$\left(\frac{c_3}{2} + \frac{c_5 h^2}{4}\right)x^2 + \left(c_1 + \frac{c_4 h^2}{8} + \frac{c_6 h^4}{64}\right) = 0 \qquad \text{(e)}$$

This is of form A·B+C-0, where A, B, C are independent. Thus,

$$c_3/2 + c_5 h^2/4 = 0 \qquad \text{(f)}$$

$$c_1 + c_4 h^2/8 + c_6 h^4/64 = 0 \qquad \text{(g)}$$

Multiply Eq. (f) by h and subtract it from Eq. (d) to find

$$c_5 = -3p/Lt$$

Then, Eqs. (g) and (a) give

$$c_3 = 3p/2htL, \qquad c_6 = 4p/th^3L$$

On $x = 0$: $V = 0$:

$$\int_{-h/2}^{h/2}\tau_{xy}t\,dy = \int_{-h/2}^{h/2}\left(-c_1 - \frac{c_4 y^2}{2} - \frac{c_6 y^4}{4}\right)t\,dy = 0$$

or

$$c_1 + c_4(h^2/24) + c_6(h^4/320) = 0 \qquad \text{(h)}$$

Subtracting Eq. (h) from (g), together with the value of c_6 already obtained, we have

$$c_4 = -3p/5htL$$

Finally, substitute c_4 and c_6 into Eq. (g) to determine

$$c_1 = ph/80tL$$

The stress function is thus,

$$\Phi = \frac{p}{Lt}\left[\frac{h}{80}xy - \frac{x^3}{12} - \frac{x^3 y}{4h} + \frac{xy^3}{10h} + \frac{x^3y^3}{h^3} - \frac{xy^5}{5h^3}\right]$$

and the stresses are

$$\sigma_x = \frac{p}{Lt}\left[\frac{3}{5h}xy + \frac{2}{h^3}x^3 y - \frac{4}{h^3}xy^3\right]$$

$$\sigma_y = \frac{p}{Lt}\left[-\frac{x}{2} - \frac{3}{2h}xy + \frac{2}{h^3}xy^3\right] \quad ◄$$

$$\tau_{xy} = \frac{p}{Lt}\left[\frac{h}{80} - \frac{3x^2}{4h} + \frac{3y^2}{10h} + \frac{3}{h^3}x^2y^2 - \frac{1}{h^3}y^4\right]$$

(b)

$$\sigma_x = \frac{Mc}{I} = \frac{(px^2/6L)(h/2)}{th^3/12} = \frac{px^2}{Lth^2} \quad ◄$$

(c) The maximum stresses are

$$(\sigma_x)_{elast.} = \frac{p}{10th}\left[6y + 2000y - \frac{40y^3}{h^2}\right]$$

$$(\sigma_x)_{elem.} = \frac{p(10h)^3}{10h^3 t} = 100(p/t) \quad ◄$$

$(\sigma_x)_{elast.} = 99.8(p/t)$ at $y = \pm h/2$

Thus,

$$(\sigma_x)_{elast.} = 0.998(\sigma_x)_{elem.} \quad ◄$$

5.9

(a) We can show that given Φ satisfies $\nabla^4\Phi=0$. From Eqs. (3.13):

$$\sigma_x = \frac{p}{0.43}\left\{2\left[0.78-\tan^{-1}\frac{y}{x}\right]-\frac{2xy}{x^2+y^2}\right\} \quad (a)$$

$$\sigma_y = \frac{p}{0.43}\left\{2\left[0.78-\tan^{-1}\frac{y}{x}\right]-2+\frac{2xy}{x^2+y^2}\right\} \quad (b)$$

$$\tau_{xy} = -\frac{p}{0.43}\frac{2y^2}{x^2+y^2} \quad (c)$$

To test which stress is maximum we rewrite σ_y in the form:

$$\sigma_y = \frac{p}{0.43}\left\{2\left[0.78-\tan^{-1}\frac{y}{x}\right]-\frac{2(x^2+y^2)}{x^2+y^2}+\frac{2xy}{x^2+y^2}\right\}$$

$$=\frac{p}{0.43}\left\{2\left[0.78-\tan^{-1}\frac{y}{x}\right]-\frac{2xy}{x^2+y^2}-\frac{2(x-y)^2}{x^2+y^2}\right\}$$

Comparing this with Eq. (a), noting $2(x-y)^2/(x^2+y^2)>0$, we conclude that: $\sigma_x > \sigma_y$.
When y=0, Eqs.(a) to (c) yield
$\sigma_x=3.63p, \sigma_y=-p/0.43, \tau_{xy}=-p/0.43$
Maximum stress occurs at y=0:

$\sigma_{x,max}=3.63(2)=7.26$ MPa ◄

(b) $(\sigma_x)_{elem.}=\frac{Mc}{I}=\frac{\frac{1}{2}px^2(\frac{x}{2})}{x^3/3}=3p=6$ MPa

Thus, $(\sigma_x)_{elast.}=1.21(\sigma_x)_{elem.}$ ◄

5.10

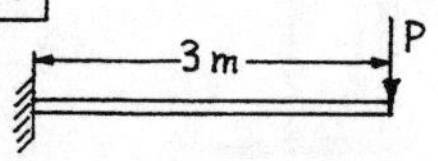

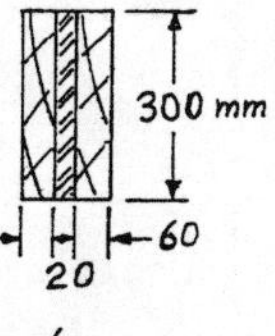

M=3P
$n=E_s/E_t=20$

$I_t=(520)(300)^3/12=1170(10^6)$ mm^4

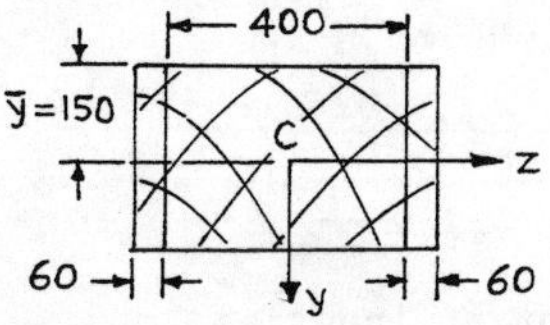

The allowable stress in the transformed section is
(120/20)=6 MPa < 7 MPa
Thus, the stress in the steel is the contralling stress. Hence,

$M_{max}=3P_{max}=\sigma_{max}I_t/c$

$=6(10^6)1170(10^{-6})/(0.15)$

or

$P_{max}=15.6$ kN ◄

5.11

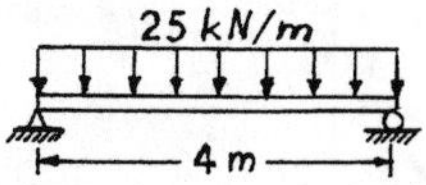

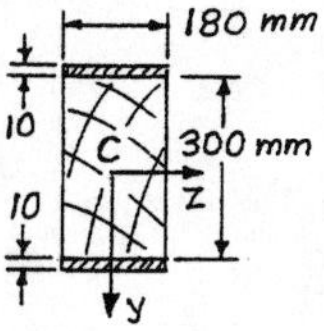

$n=E_a/E_w=7$

$M_{max}=pL^2/8=25\times10^3(4)^2/8=50$ kN·m

$I_t=\frac{1}{12}(180)(300)^3+$
$2[\frac{1260(10^3)}{12}+1260)(10)(155)^2]$
$=1010.64(10^6)$ mm^4

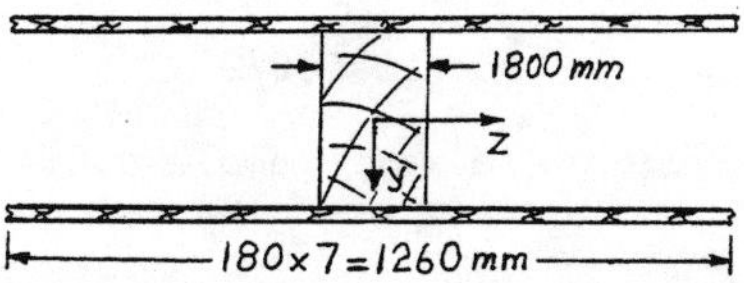

$\sigma_{w,max}=\frac{Mc}{I_t}=\frac{50\times10^3(0.15)}{1010.64(10^{-6})}=7.4$ MPa ◄

$\sigma_{a,max}=\frac{7Mc}{I_t}=\frac{7\times50\times10^3(0.16)}{1010.64(10^{-6})}$

$=55.4$ MPa ◄

5.12

Equation (5.54) becomes
$(kd)^2+(kd)(20/300)(1200)-(20/300)(500)(1200)=0$
or
$(kd)^2+180kd-40(10^3)=0$
Solving,
kd=164 mm
Hence,
500-kd=336 mm

From Eqs. (e) of Example 5.5:

$M_c=\frac{1}{2}\sigma_c(bkd)(d-kd/3)$

$=\frac{1}{2}(12\times10^6)(0.3\times0.164)(0.5-\frac{0.164}{3})$

$=89.5$ kN·m

and

$M_s=\sigma_sA_s(d-kd/3)$

$=150(10^6)(1200\times10^{-6})(0.445)$

$=81.9$ kN·m

Thus,

$M_{all}=81.9$ MPa ◄

5.13

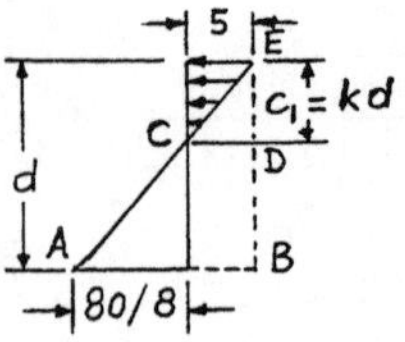

The stresses in concrete and the equivalent of the steel (Fig. 5.14b) have the values shown in the preceding figure. From the similarity of ΔECD and ΔEAB, we find

$$\frac{c_1}{d} = \frac{5}{(80/8)+5} = \frac{5}{15}$$

or

$$c_1 = 5d/15 = 500/3$$

Equation (d) of Example 5.5:

$$\frac{1}{2}\sigma_c(c_1 b) = \sigma_s A_s$$

$$\frac{1}{2}(5)(\frac{500}{3})300 = 80A_s$$

or

$$A_s = 1563 \text{ mm}^2$$ ◄

Then, we have from Eq. (e) of Example 5.5,

$$M = \sigma_s A_s(d - c_1/3)$$

$$= 80(10^6)(1563\times10^{-6})(0.5 - \frac{0.5}{9})$$

$$= 55.6 \text{ kN·m}$$ ◄

5.14

The shearing stress at a distance s is given by

$$\tau = \frac{V_y Q_z}{I_z b} = \frac{V_y}{I_z}\int_0^{\alpha} R\cos\theta(t\,R\,d\theta)$$

$$= \frac{2V_y}{\pi R t}\sin\alpha$$

This shows that $\tau = 0$ at the free ends and τ_{max} at the neutral axis, same as for a rectangular section.

The shearing stress produces the following twisting moment about O:

$$T = \int \tau R\, dA = \int_0^{\pi} \frac{2V_y \sin\alpha}{\pi R t} R(Rt\, d\alpha)$$

$$= 4RV_y/\pi$$

By applying the principle of moments at O: $V_y e = M$. Thus,

$$e = 4R/\pi$$ ◄

5.15

Shearing stress in the web is neglected. Moment of the forces about S:

$$V_1 e_1 = V_2 e_2; \quad \frac{e_1}{e_2} = \frac{V_2}{V_1} \quad \text{(a)}$$

Let M_1 and M_2 be bending moments on flanges 1 and 2, respectively. Curvature-moment are related by

$$\frac{1}{r_1} = \frac{M_1}{EI_1}, \quad \frac{1}{r_2} = \frac{M_2}{EI_2} \quad \text{(b)}$$

By assuming $r_1 = r_2$, we have

$$\frac{M_1}{EI_1} = \frac{M_2}{EI_2}; \quad \frac{dM_1}{I_1} = \frac{dM_2}{I_2} \quad \text{(c)}$$

Introducing $dM/dx = -V$, Eq. (c) gives

$$V_1/V_2 = I_1/I_2$$

Equation (a) now becomes

$$e_1/e_2 = I_2/I_1$$

Since

$$e_1 + e_2 = h; \quad [e_1/(h - e_1)] = I_2/I_1$$

Thus,

$$e_1 = I_2 h/(I_1 + I_2)$$ ◄

where

$$I_1 = b_1^3 t_1/12 \qquad I_2 = b_2^3 t_2/12$$

5.16

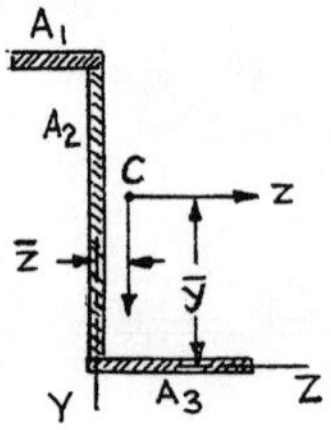

Fig. (a)

Location of centroid C (Fig. a):

$$\bar{z} = \frac{15.5\times75(-37.5) + 0 + 9.5\times125\,(62.5)}{15.5\times75 + 9.5\times250 + 9.5\times125}$$

$$= 6.481 \text{ mm}$$

$$\bar{y} = \frac{15.5\times75(-250) + 0 + 9.5\times125\,(-125)}{15.5\times75 + 9.5\times250 + 9.5\times125}$$

$$= -124.339 \text{ mm}$$

Moment of inertia:

$$I_y = \frac{1}{12}(15.5)(75)^3 + (15.5\times75)(43.98)^2 + \frac{1}{12}(250)(9.5)^3 + 250(9.5)(6.48)^2 + \frac{1}{12}(9.5)(125)^3 + 9.5(125)(56.01)^2$$

$$= 8.18(10^6) \text{ mm}^4$$

(CONT.)

5.16 CONT.

$I_z=75(15.5)^3/12+(75\times15.5)(125.6)^2+9.5(250)^3/12+(9.5\times250)(0.65)^2+125(9.5)^3/12+(125\times9.5)(124.3)^2=49.10(10^6)\ \text{mm}^4$

$I_{yz}=(15.5\times75)(-125)(-43.98)+(9.5\times250)(-0.65)(-6.48)+(9.5\times125)(124.3)(56)=14.70(10^6)\ \text{mm}^4$

Then,

$$\theta_p=\frac{1}{2}\tan^{-1}\left[-\frac{2(14.7)}{8.18-49.1}\right]=17.85°$$

$$I_{y'}=10^6\left[\frac{8.18+49.1}{2}+\frac{8.18-49.1}{2}\cos35.7° -14.7\sin35.7°\right]=3.45(10^6)\ \text{mm}^4$$

Similarly, we compute

$I_1=I_{z'}=54.1(10^6)\ \text{mm}^4$

$I_2=I_{y'}=3.45(10^6)\ \text{mm}^4$

From geometry of section (Fig. b):

HB=144.57 mm BC=39.05 mm

Thus,

$$\tau_{xz}=\frac{V_{y'}}{I_{z'}t}[st(0.1445-\frac{s}{2}\sin17.85°]ds$$

Shear force due to $V_{y'}$ (Fig. b):

$$F_1=\int_0^s\tau_{xz}t\,ds=\frac{V_{y'}(0.0155)}{54.1\times10^{-6}}\int_0^{0.075}\left[0.1445s-\frac{s^2}{2}\sin17.85°\right]ds$$

$=0.1108V_{y'}$

We write

$V_{y'}\cdot e_{z'}=0.25F_1=0.25(0.1108V_{y'})$

or

$e_{z'}=0.0277\ \text{m}=27.7\ \text{mm}$ ◄

Shear force due to $V_{z'}$. Now assume that the direction of F_1 shown in the figure is reversed. Then,

$$F_1=\frac{V_{z'}t}{I_{y'}}\int_0^{0.075}\left[0.03905s-\frac{s^2}{2}\cos17.85°\right]ds$$

$=0.1928V_{z'}$

We write

$V_{z'}\cdot e_{y'}=0.25(0.1928V_{z'})$

or

$e_{y'}=0.0482\ \text{m}=48.2\ \text{mm}$ ◄

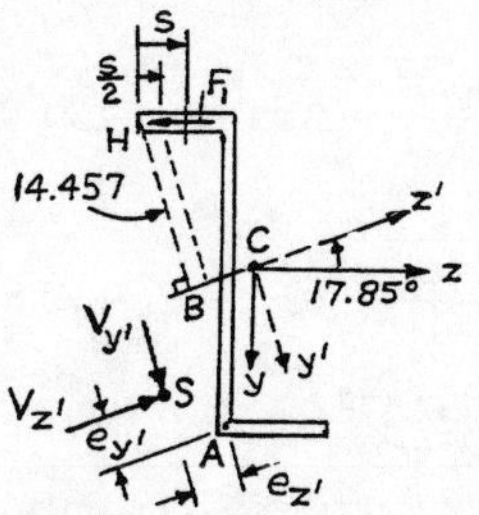

Fig. (b)

5.17

$$\tau_{max}=\frac{VQ}{Ib}=\frac{pL}{2}\frac{(th/2)(h/4)}{(th^3/12)t}=\frac{3}{4}\frac{pL}{ht}$$

$$\sigma_{max}=\frac{Mc}{I}=\frac{pL^2}{8}\frac{h/2}{th^3/12}=\frac{3}{4}\frac{pL^2}{th^2}$$

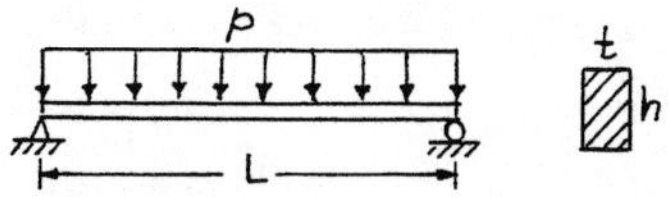

Thus,

$\sigma_{max}/\tau_{max}=L/h$

from which

$$L=\frac{\sigma_{max}h}{\tau_{max}}=\frac{8.4(0.15)}{0.7}=1.8\ \text{m}$$ ◄

Then,

$$p=\frac{4}{3}\frac{\tau_{max}(th)}{L}=\frac{4}{3}\frac{700(0.05\times0.15)}{1.8}$$

$=3.88$ kN·m ◄

5.18

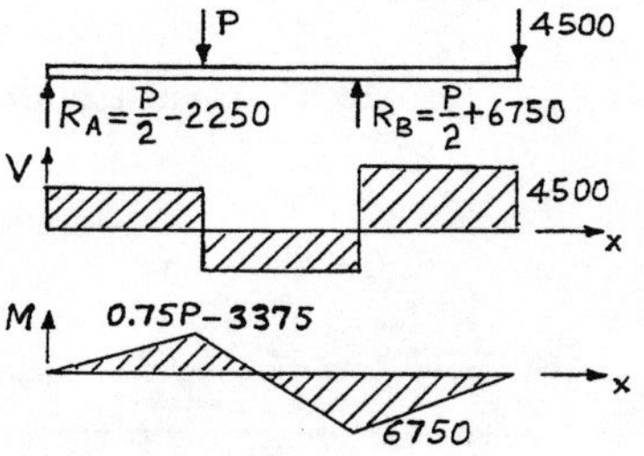

We have

$$I=\frac{0.2(0.25)^3}{12}-\frac{0.15(0.2)^3}{12}=160.417(10^{-6})\ \text{m}^4$$

Then,

$$\tau=\frac{VQ}{Ib}=0.7(10^6)=\frac{(P/2)+2250}{160.417\times10^{-6}(0.05)}(0.2\times0.025\times0.1125+0.1\times0.025\times0.05)$$

Solving,

P=9.32 kN

Similarly,

$$\sigma=\frac{Mc}{I}=7(10^6)=\frac{(0.75P-3375)(0.125)}{160.417\times10^{-6}}$$

or

P=16.478 kN

Thus,

$P_{all}=9.32$ kN ◄

5.19

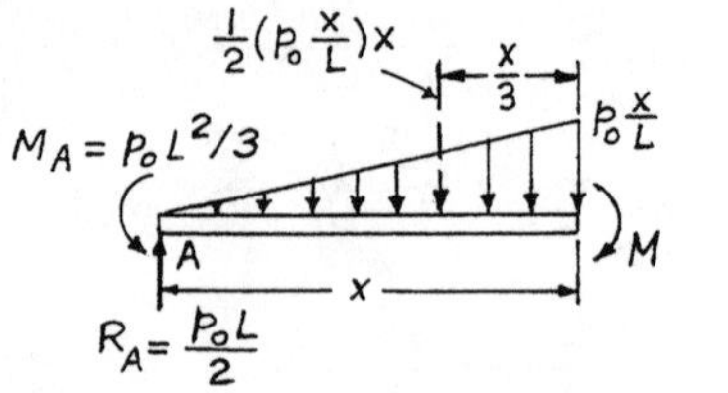

We have

$$EI\upsilon'' = M = \frac{1}{6}p_o x^3 - \frac{1}{2}p_o Lx + \frac{1}{3}p_o L^2$$

Integrate

$$EI\upsilon' = \frac{p_o}{24L}x^4 - \frac{p_oL}{4}x^2 + \frac{p_oL^2}{3}x + c_1$$

$\upsilon'(0)=0;\quad c_1=0.$

$$EI\upsilon' = \frac{p_o}{24L}x^4 - \frac{p_oL}{4}x^2 + \frac{p_oL^2}{3}x \qquad \text{(a)}$$

Integrate

$$EI\upsilon = \frac{p_o}{120EIL}x^5 - \frac{p_oL}{12}x^3 + \frac{p_oL^2}{6}x^2 + c_2$$

$\upsilon(0)=0;\quad c_2=0.$

(a) $$\upsilon = \frac{p_o}{120EIL}(x^5 - 10L^2x^3 + 20L^3x^2) \qquad ◄$$

(b) Let x=L in this equation:

$$\upsilon_B = \frac{11p_oL^4}{120EI} \qquad ◄$$

(c) Let x=L in Eq. (a):

$$\theta_B = \frac{p_oL^3}{8EI} \qquad ◄$$

5.20

$$EI\upsilon^{IV} = p \qquad EI\upsilon''' = px + c_1$$

$$EI\upsilon'' = \frac{1}{2}px^2 + c_1x + c_2$$

$$EI\upsilon' = \frac{1}{6}px^3 + \frac{1}{2}c_1x^2 + c_2x + c_3$$

$$EI\upsilon = \frac{1}{24}px^4 + \frac{1}{6}c_1x^3 + \frac{1}{2}c_2x^2 + c_3x + c_4$$

Boundary conditions:

$EI\upsilon(0)=0;\quad c_4=0$

$EI\upsilon''(0)=0;\quad c_2=0$

$$EI\upsilon(L)=0;\quad \frac{pL^3}{24} + \frac{c_1L^2}{6} + c_3 = 0 \qquad \text{(a)}$$

$$EI\upsilon'(L) = -EI\frac{pL^3}{96EI} = \frac{pL^3}{6} + \frac{c_1L^2}{2} + c_3$$

or $\quad c_3 = -17pL^3/96 - c_1L^3/2 \qquad$ (b)

Substituting Eq. (b) into (a), we obtain reaction at right end:

$$c_1 = -13pL/32 = R \qquad ◄$$

5.21

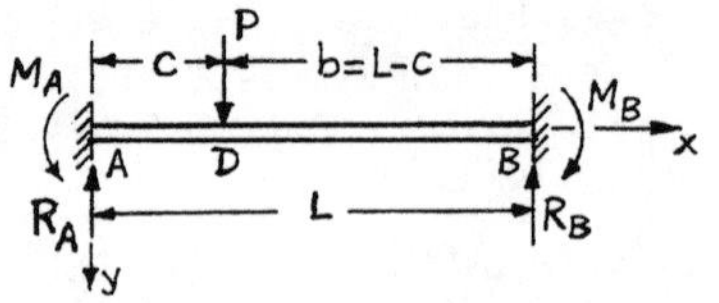

Segment AD:

$$EI\upsilon_1'' = M_A - R_Ax$$

$$EI\upsilon_1' = M_Ax - \frac{1}{2}R_Ax^2 + c_1$$

$$EI\upsilon_1 = \frac{1}{2}M_Ax^2 - \frac{1}{6}R_Ax^3 + c_1x + c_2 \qquad \text{(a)}$$

Segment BD:

$$EI\upsilon_2'' = M_A - R_Ax + P(x-c)$$

$$EI\upsilon_2' = M_Ax - \frac{1}{2}R_Ax^2 + \frac{1}{2}P(x-c)^2 + c_3$$

$$EI\upsilon_2 = \frac{1}{2}M_Ax^2 - \frac{1}{6}R_Ax^3 + \frac{1}{6}P(x-c)^3 + c_3x + c_4 \qquad \text{(b)}$$

Boundary conditions:

$\upsilon_1(0)=0;\quad c_2=0$

$\upsilon_1'(0)=0;\quad c_1=0$

$\upsilon_1(c)=\upsilon_2(c);\quad c_3c+c_4=0$

$\upsilon_1'(c)=\upsilon_2'(c);\quad c_3=0,\quad c_4=0$

and

$$\upsilon_2'(L) = 0 = M_AL - \frac{1}{2}R_AL^2 + \frac{1}{2}Pb^2 \qquad \text{(c)}$$

$$\upsilon_2(L) = 0 = \frac{1}{2}M_AL^2 - \frac{1}{6}R_AL^3 + \frac{1}{6}Pb^3 \qquad \text{(d)}$$

Solving Eqs. (c) and (d),

$$R_A = \frac{Pb^3}{L^3}(3c+b) = \frac{P(L-c)^2}{L^3}(2c+L) \qquad ◄$$

Substitution of this into Eq. (c) gives

$$M_A = \frac{Pcb^2}{L^2} = \frac{Pc(L-c)^2}{L^2} \qquad ◄$$

Thus, introducing the values of M_A, R_A, and $c_1=c_2=c_3=c_4=0$ into Eqs. (a) and (b) we obtain the deflections.

For $0 \le x \le c$:

$$\upsilon_1 = \frac{P(L-c)^2x^2}{6EIL^3}(3cL-2cx-Lx)$$

For $c \le x \le L$:

$$\upsilon_2 = \frac{P(L-c)^2x^2}{6EIL^3}(3cL-2cx-Lx) + \frac{P}{6EI}(x-c)^3$$

5.22

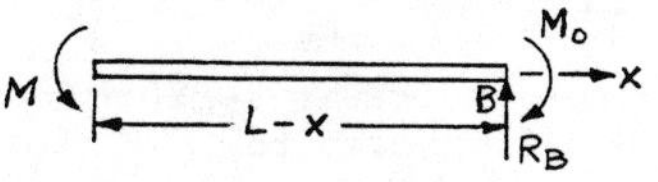

The reactions are statically indeterminate. We have

$$EI\,v'' = M = R_B x - R_B L + M_o$$

$$EI\,v' = \tfrac{1}{2}R_B x^2 - R_B L x + M_o x + c_1$$
$$v'(0) = 0; \qquad c_1 = 0.$$

$$EI\,v = \tfrac{1}{6}R_B x^3 - \tfrac{1}{2}R_B L x^2 + \tfrac{1}{2}M_o x^2 + c_2$$
$$v(0) = 0; \qquad c_2 = 0$$
$$v(L) = 0; \qquad R_B = 3M_o/2L \qquad ◄$$

The preceding equation gives

$$v = M_o x^2 (x - L)/4EIL \qquad ◄$$

5.23

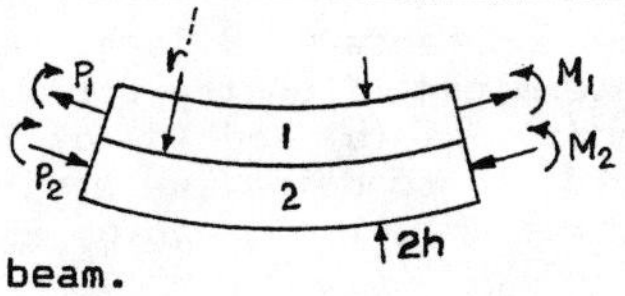

Bent beam.

(a) Static equilibrium gives

$$P_1 = P_2 = P \qquad M_1 + M_2 = Ph$$

Equation (5.9):

$$M_1 = E_1 I_1/r \qquad M_2 = E_2 I_2/r$$

Interface strains must be the same:

$$\underbrace{\alpha_1 \Delta T}_{\text{due to temp. increase}} + \underbrace{\frac{P_1}{E_1 h}}_{\text{due to axial force}} + \underbrace{\frac{h/2}{r}}_{\text{due to bending (from Eq. 5.9)}} = \alpha_2 \Delta T - \frac{P_2}{E_2 h} - \frac{h/2}{r}$$

This yields an expression for the interface curvature:

$$\frac{1}{r} = \frac{12(\alpha_2 - \alpha_1)\Delta T}{h(14 + n + 1/n)} \qquad ◄$$

where $n = E_1/E_2$.

(b) At the interface

$$\sigma_1 = \frac{P}{h} + \frac{E_1 h}{2r} = (\sigma_{yP})_1$$
$$\sigma_2 = -\frac{P}{h} - \frac{E_2 h}{2r} = -(\sigma_{yP})_2 \qquad ◄$$

(c) Summing these equations,

$$(\sigma_{yP})_1 - (\sigma_{yP})_2 = -h(E_1 - E_2)/2r$$

It follows that

$$\frac{(\sigma_{yP})_1 - (\sigma_{yP})_2}{E_1 - E_2} = \frac{h}{2}\,\frac{24(\alpha_2 - \alpha_1)\Delta T}{h(14 + n + 1/n)}$$

from which

$$\Delta T = \frac{14 + n + 1/n}{6(\alpha_2 - \alpha_1)}\,\frac{(\sigma_{yP})_1 - (\sigma_{yP})_2}{E_1 - E_2} \qquad ◄$$

5.24

B

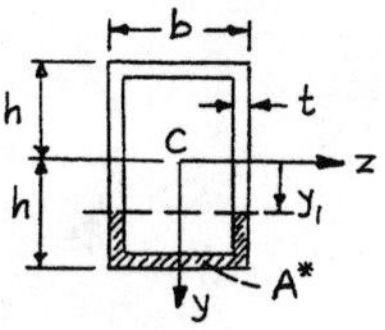

$$Q = Q_{outside} - Q_{inside}$$
$$= \frac{b}{2}(h^2 - y_1^2) - \frac{b-2t}{2}[(h-t)^2 - y_1^2]$$
$$= bht - \frac{1}{2}bt^2 + th^2 + t^3 - ty_1^2$$

$$\alpha = \frac{A}{I^2}\int \frac{Q}{(width)^2}\,dA$$

$$= \frac{A}{I^2}\left\{2\left[\int_0^{h-t} \frac{(bht + th^2 - \frac{bt^2}{2} - 2ht^2 + t^3 - ty_1^2)^2}{b^2} 2t\,dy_1 + \int_{h-t}^{h} \frac{(bht - \frac{bt^2}{2} + th^2 - 2ht^2 + t^3 - ty_1^2)^2}{b^2} b\,dy_1\right]\right\}$$

or

$$\alpha = \frac{A}{A_{web}} = 1 + \frac{b}{2(h-t)}$$

where

$$A = 2bh - (b-2t)(2h-2t) = 2bt + 4th - 4t^2$$
$$A_{web} = 2(2h-2t)t = 4ht - 4t^2$$
$$I = \tfrac{2}{3}bh^3 - \tfrac{2}{3}(b-2t)(h-t)^3$$

B

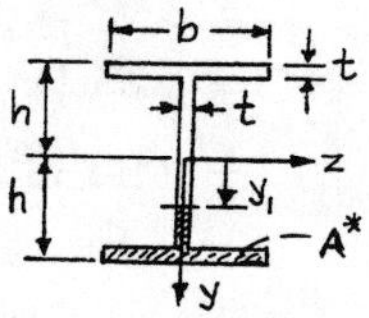

$$Q = (h - t - y_1)t[y_1 + \tfrac{1}{2}(h - t - y_1)]$$

$$= \frac{th^2}{2} - ht^2 + \frac{t^3}{2} + bht - \frac{bt^2}{2} - \frac{t}{2}y_1^2$$

$$\alpha = \frac{A}{I^2}\left\{2\left[\int_0^{h-t} \frac{(bht - ht^2 + \frac{th^2}{2} - \frac{bt^2}{2} - \frac{ty_1^2}{2} + \frac{t^3}{2})^2}{t^2} t\,dy_1 + \int_{h-t}^{h} \frac{(bht - ht^2 + \frac{th^2}{2} - \frac{bt^2}{2} - \frac{ty_1^2}{2} + \frac{t^3}{2})^2}{b^2} b\,dy_1\right]\right\}$$

or

$$\alpha = \frac{A}{A_{web}} = 1 + \frac{b}{h-t} \qquad ◄$$

where

$$A = 2bt + (2h-2t)t = 2bt + 2ht - 2t^2$$
$$A_{web} = (2h-2t)t = 2ht - 2t^2$$
$$I = \tfrac{1}{12}t(2h-2t)^3 + 2[\tfrac{1}{12}bt^3 + bt(h - \tfrac{t}{2})^2]$$

Note: For thin-walled sections flange and web thicknesses are small with compared to unity and their products are neglected.

(CONT.)

5.24 CONT.

C

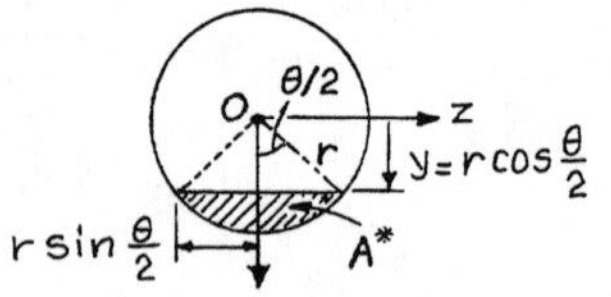

We write

$$A=\frac{r^2}{2}(\theta-\sin\theta)=\text{area of segment}$$

$$dA=\frac{r^2}{2}(1-\cos\theta)d\theta=r^2(1-\cos^2\frac{\theta}{2})d\theta$$

$$Q=\int_{A^*} y\ dA=\int_{A^*}(r\cos\frac{\theta}{2})r^2(1-\cos^2\frac{\theta}{2})d\theta$$

$$=r^3\int_0^{\theta}(\cos\frac{\theta}{2}-\cos^3\frac{\theta}{2})d\theta$$

$$=r^3[2\sin\frac{\theta}{2}-\frac{2}{3}\sin\frac{\theta}{2}(\cos^2\frac{\theta}{2}+2)]$$

$$=\frac{2}{3}r^3[\sin\frac{\theta}{2}-\sin\frac{\theta}{2}\cos^2\frac{\theta}{2}]$$

$$\alpha=\frac{A}{I^2}\int\frac{Q^2}{(width)^2}dA$$

$$=\frac{A}{I^2}\int_0^{2\pi}\frac{[2\pi r^3/3(\sin\theta/2-\sin\theta/2\cos^2\theta/2]^2}{(2r\sin\theta/2)^2}r^2(1-\cos^2\frac{\theta}{2})d\theta$$

Since,

$$A=\pi r^2 \qquad I^2=\pi^2 r^8/16$$

we have, after simplification:

$$\alpha=\frac{16}{\pi r^6}\int_0^{2\pi}(\frac{r^6}{9})(1-\cos^2\frac{\theta}{2})^3 d\theta$$

$$=10/9$$ ◄

D

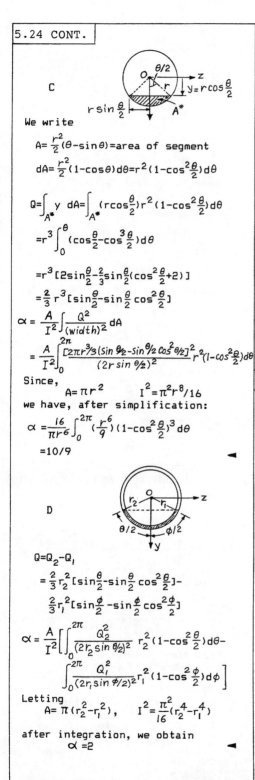

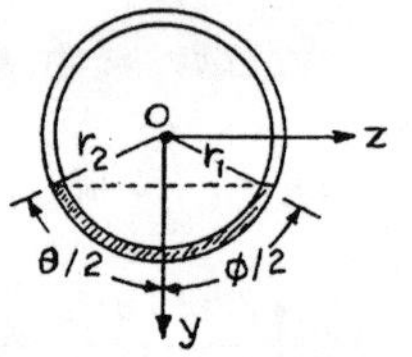

$$Q=Q_2-Q_1$$

$$=\frac{2}{3}r_2^2[\sin\frac{\theta}{2}-\sin\frac{\theta}{2}\cos^2\frac{\theta}{2}]-$$

$$\frac{2}{3}r_1^2[\sin\frac{\phi}{2}-\sin\frac{\phi}{2}\cos^2\frac{\phi}{2}]$$

$$\alpha=\frac{A}{I^2}\left[\int_0^{2\pi}\frac{Q_2^2}{(2r_2\sin\theta/2)^2}r_2^2(1-\cos^2\frac{\theta}{2})d\theta-\int_0^{2\pi}\frac{Q_1^2}{(2r_1\sin\phi/2)^2}r_1^2(1-\cos^2\frac{\phi}{2})d\phi\right]$$

Letting

$$A=\pi(r_2^2-r_1^2), \qquad I^2=\frac{\pi^2}{16}(r_2^4-r_1^4)$$

after integration, we obtain

$$\alpha=2$$ ◄

5.25

Since stress is symmetrical, Eq. (3.37) reduces to

$$(\frac{\partial^2}{\partial r^2}+\frac{1}{r}\frac{\partial}{\partial r})(\frac{\partial^2\Phi}{\partial r^2}+\frac{1}{r}\frac{\partial\Phi}{\partial r})=$$

$$\frac{\partial^4\Phi}{\partial r^4}+\frac{2}{r}\frac{\partial^3\Phi}{\partial r^3}-\frac{1}{r^2}\frac{\partial^2\Phi}{\partial r^2}+\frac{1}{r^3}\frac{\partial\Phi}{\partial r}=0$$

The given Φ satisfies this equation.

Equations (3.29) leads to

$$\sigma_r=\frac{1}{r}\frac{\partial\Phi}{\partial r}=\frac{A}{r^2}+B(1+2\cdot\ln r)+2C$$ ◄

$$\sigma_\theta=\frac{\partial^2\Phi}{\partial r^2}=(-\frac{A}{r^2})+B(3+2\cdot\ln r)+2C$$

The constants A, B, and C are determined from the boundary conditions (b) and (d) of Sec. 5.13. In so doing, we arrive at the solution given by Eq. (5.67).

5.26

From Eq. (5.68), we have

$$Z=\frac{1}{A}\left\{\int_{-c}^{-c_2}\frac{y}{R+y}b\,dy+\int_{-c_2}^{-c_1}\frac{y}{R+y}t\,dy\right\}$$

$$=-\frac{1}{A}\left\{b\left[\int_{-c}^{-c_2}\left(1-\frac{R}{R+y}\right)dy+t\left[\int_{-c_2}^{-c_1}\left(1-\frac{R}{R+y}\right)dy\right]\right\}$$

This expression, after integration gives

$$Z=-\frac{1}{A}\Big[b(c-c_2)-Rb\cdot\ln\frac{R-c_2}{R-c}+$$

$$t(c_1+c_2)-Rt\cdot\ln\frac{R+c_1}{R-c}\Big]$$

$$=-1+\frac{R}{A}\Big[t\cdot\ln(R+c_1)+$$

$$(b-t)\cdot\ln(R-c_2)-b\cdot\ln(R-c)\Big]$$ ◄

Here the area of the cross section of the beam is

$$A=c_1t+cb-c_2(b-t)$$

5.27

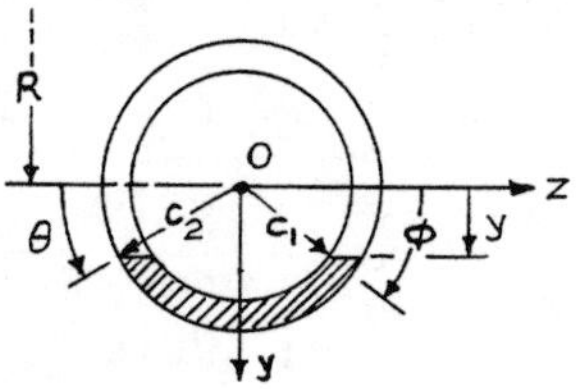

Let $\alpha_1=R/c_1$ and $\alpha_2=R/c_2$.
Referring to the preceding figure:

$$A=\pi(c_2^2-c_1^2)$$

and

$$dA_1=2c_1\cos\phi\,dy$$
$$dA_2=2c_2\cos\theta\,dy$$
$$y=c_1\sin\phi=c_2\sin\theta$$
$$dy=c_1\cos\phi\,d\phi=c_2\cos\theta\,d\theta$$

Applying Eq. (5.68) :

$$Z=-\frac{1}{A}\int\frac{y}{R+y}\,dA$$

$$=-\frac{1}{\pi(c_2^2-c_1^2)}\left\{\int_{-\pi/2}^{\pi/2}\frac{c_2\sin\theta}{R+c_2\sin\theta}\,2c_2^2\cos^2\theta\,d\theta-\int_{-\pi/2}^{\pi/2}\frac{c_1\sin\phi}{R+c_1\sin\phi}\,2c_1^2\cos^2\phi\,d\phi\right\}$$

$$=-\frac{2}{\pi(c_2^2-c_1^2)}\left\{\int_{-\pi/2}^{\pi/2}c_2^2\,\frac{\sin\theta-\sin^3\theta}{\sin\theta+\alpha_2}\,d\theta-\int_{-\pi/2}^{\pi/2}c_1^2\,\frac{\sin\phi-\sin^3\phi}{\sin\phi+\alpha_1}\,d\phi\right\}$$

$$=-\frac{2}{\pi(c_2^2-c_1^2)}\left\{\int_{-\pi/2}^{\pi/2}c_2^2\left[\sin^2\theta-\alpha_2\sin\theta+(\alpha_2^2-1)-\frac{\alpha_2(\alpha_2^2-1)}{\sin\theta+\alpha_2}\right]d\theta-\int_{-\pi/2}^{\pi/2}c_1^2\left[\sin^2\phi-\alpha_1\sin\phi+(\alpha_1^2-1)-\frac{\alpha_1(\alpha_1^2-1)}{\sin\phi+\alpha_1}\right]d\phi\right\}$$

Integrating and simplifying this leads to

$$Z=1-2+\frac{2c_2^2\alpha_2^2-2c_1^2\alpha_1^2}{c_2^2-c_1^2}-\frac{2c_2^2\alpha_2\sqrt{\alpha_2^2-1}-2c_1^2\alpha_1\sqrt{\alpha_1^2-1}}{c_2^2-c_1^2}$$

$$=-1+\frac{2Rc_1}{c_2^2-c_1^2}\sqrt{\frac{R^2}{c_1^2}-1}-\frac{2Rc_2}{c_2^2-c_1^2}\sqrt{\frac{R^2}{c_2^2}-1}$$

or

$$Z=-1+\frac{2R}{c_2^2-c_1^2}\left(\sqrt{R^2-c_1^2}-\sqrt{R^2-c_2^2}\right)$$ ◄

5.28

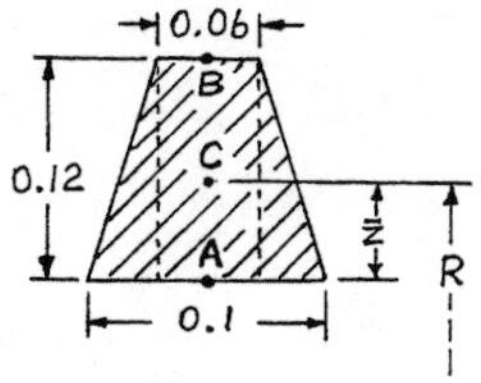

We compute:

$$A=(0.06)(0.12)+2[\tfrac{1}{2}(0.02)(0.12)]=0.0096\ \text{m}^2$$

$$\bar{z}=\frac{\sum A_i z_i}{\sum A_i}=0.055\ \text{m},\quad R=0.155\ \text{m}$$

$h=0.12$ m, $b_1=0.06$ m, $c=0.055$ m
$c_1=0.065$ m, $M=1.55P$, $b=0.1$ m
Expression of Fig. C (Table 5.3) yields then

$$Z=-1+\frac{0.155}{(0.0096)(0.12)}\left\{[0.06(0.12)+(0.155+0.065)(0.1-0.06)]\cdot\ln\frac{0.22}{0.1}-0.4(0.12)\right\}$$

$$=0.05154$$ ◄

At point A:

$$-120(10^6)=-\frac{P(0.055)}{0.0515(0.0096)(0.155-0.055)}$$

or

$$P=108\ \text{kN}=P_{all}$$ ◄

At point B:

$$120(10^6)=\frac{P(0.065)}{0.0515(0.0096)(0.155+0.065)}$$

or

$$P=201\ \text{kN}$$

5.29

$$Z=-1+2\left(\frac{75}{25}\right)^2-\left(\frac{75}{25}\right)\sqrt{\left(\frac{75}{25}\right)^2-1}=0.0294$$

(a) Equation (5.71):

$$161(10^6)=\frac{P}{\pi(0.025)^2}+\frac{-0.0125P}{\pi(0.025)^2(0.075)}\times\left[1+\frac{-0.025}{0.0294(0.075-0.025)}\right]$$

or

$$P=11.436\ \text{kN}$$ ◄

(b)

$$(\sigma_\theta)_B=\frac{11{,}436}{\pi(0.025)^2}+\frac{-0.125(11{,}436)}{\pi(0.025)^2(0.075)}\times\left[1+\frac{0.025}{0.0294(0.075+0.025)}\right]$$

$$=-86.3\ \text{MPa}$$ ◄

5.30

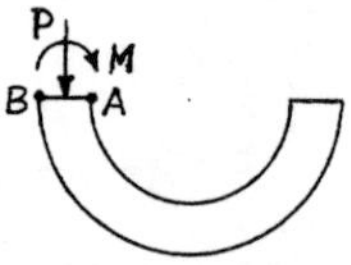

Fig. (a)

Equation (5.66) yields

$$N=(1-\frac{0.1^2}{0.2^2})^2-4\ \frac{0.1^2}{0.2^2}\ln^2\frac{0.1}{0.2}=0.082$$

$$M=PR=70(10^3)\times0.15=10.5\ \text{kN}\cdot\text{m}$$

Thus, from Eq. (5.67):

$$(\sigma_\theta)_A=\frac{4(10{,}500)}{0.05(0.2)^2(0.082)}\times$$

$$[(1-\frac{0.1^2}{0.2^2})(1+\ln1)-(1+1)\ln\frac{0.2}{0.1}]$$

$$=163\ \text{MPa}$$

$$(\sigma_\theta)_B=\frac{4(10{,}500)}{0.05(0.2)^2(0.082)}\times$$

$$[(1-\frac{0.1^2}{0.2^2})(1+\ln2)-(1+\frac{0.1^2}{0.2^2})\ln2]$$

$$=103\ \text{MPa}$$

Referrring to Fig. (a),

$$\sigma_{max}=(\sigma_\theta)_A-\frac{P}{A}=-163-\frac{70}{5}\ =-177\ \text{MPa}$$ ◄

$$\sigma_{min}=(\sigma_\theta)_B-\frac{P}{A}=103-14=89\ \text{MPa}$$

Note that Winkler's solution (from Example 5.10):

$\sigma_{max}=-176$ MPa and $\sigma_{min}=88$ MPa ◄

5.31

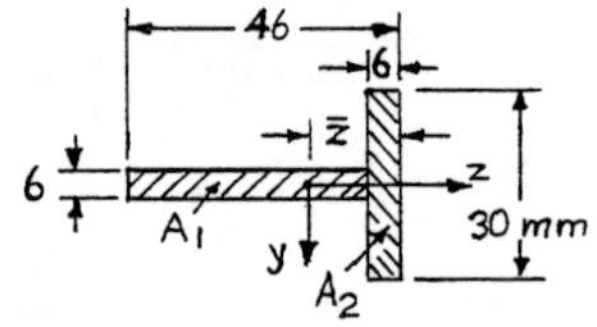

$$\bar{z}=(A_1z_1+A_2z_2)/(A_1+A_2)=16.14\ \text{mm}$$

From Table 5.3(E):

$$Z=-1+\frac{136.14}{420}[6\cdot\ln(136.14+29.86)+$$

$$(30-6)\cdot\ln(136.14-10.14)-$$

$$30\cdot\ln(136.14-16.14)]=0.009988$$

We have

$$M=-(30+120+16.14)P=-166.14P$$

Outer Edge. Applying Eq. (5.69),

(CONT.)

5.31 CONT.

$$-80\ \frac{N}{mm^2}=\frac{P}{420}+\frac{-166.14P}{420(136.14)}\Big[1+$$

$$\frac{29.86}{0.009988(136.14+29.86)}\Big]$$

or

$P=1.517$ kN$=P_{all}$ ◄

Inner Edge. Using Eq. (5.69):

$$80\ \frac{N}{mm^2}=\frac{P}{420}+\frac{166.14P}{420(136.14)}\Big[1+$$

$$\frac{-16.14}{0.009988(136.14-16.14)}\Big]$$

Solving,

$P=2.078$ kN ◄

5.32

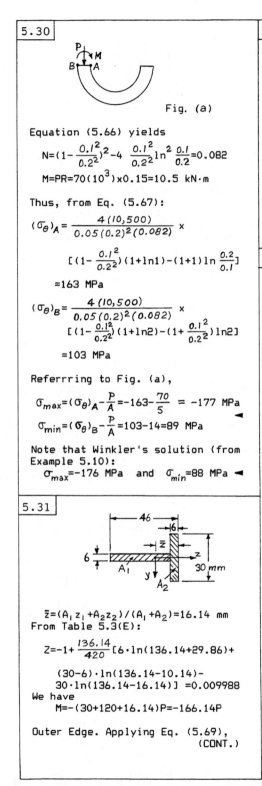

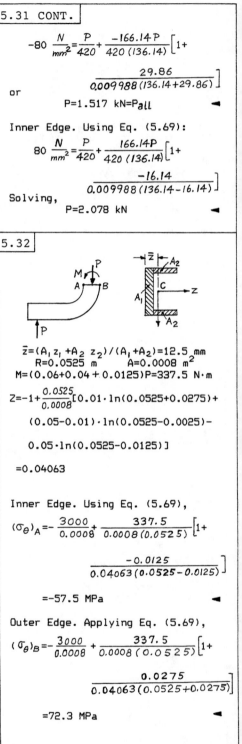

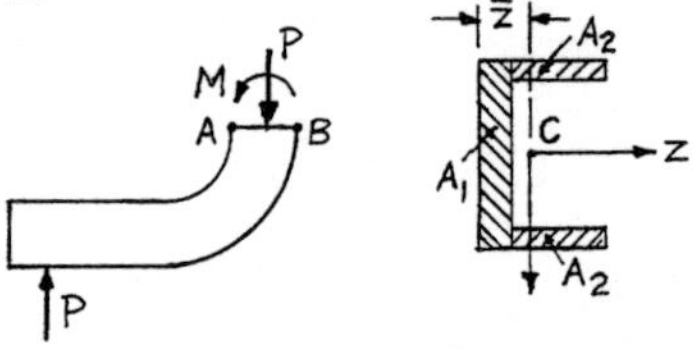

$$\bar{z}=(A_1z_1+A_2z_2)/(A_1+A_2)=12.5\ \text{mm}$$

$R=0.0525$ m $A=0.0008\ m^2$

$$M=(0.06+0.04+0.0125)P=337.5\ \text{N}\cdot\text{m}$$

$$Z=-1+\frac{0.0525}{0.0008}[0.01\cdot\ln(0.0525+0.0275)+$$

$$(0.05-0.01)\cdot\ln(0.0525-0.0025)-$$

$$0.05\cdot\ln(0.0525-0.0125)]$$

$$=0.04063$$

Inner Edge. Using Eq. (5.69),

$$(\sigma_\theta)_A=-\frac{3000}{0.0008}+\frac{337.5}{0.0008(0.0525)}\Big[1+$$

$$\frac{-0.0125}{0.04063(0.0525-0.0125)}\Big]$$

$$=-57.5\ \text{MPa}$$ ◄

Outer Edge. Applying Eq. (5.69),

$$(\sigma_\theta)_B=-\frac{3000}{0.0008}+\frac{337.5}{0.0008(0.0525)}\Big[1+$$

$$\frac{0.0275}{0.04063(0.0525+0.0275)}\Big]$$

$$=72.3\ \text{MPa}$$ ◄

5.33

Using Eq.(P5.34), at section A-B of Fig. P5.33:
M=-0.182PR=-0.182(0.75b)P=-0.136Pb

(a) Equation (5.66) yields,

$$N=(1-\frac{1}{4})^2-4(\frac{1}{4})(\ln 2)^2=0.082$$

Point A. Applying Eq. (5.67),

$$(\sigma_\theta)_A=\frac{4M}{tb^2N}[(1-\frac{1}{4})(1+\ln 1)-2\cdot\ln 2]+\frac{P}{bt}$$

$$=\frac{4(-0.136\,Pb)}{tb^2(0.082)}[\frac{3}{4}-1.3863]+\frac{P}{bt}$$

=5.22P/bt

Point B. Using Eq. (5.67),

$$(\sigma_\theta)_B=\frac{4(-0.136\,Pb)}{tb^2(0.082)}[\frac{3}{4}(1+\ln 2)-\frac{5}{4}\ln 2]+\frac{P}{bt}$$ ◄

=-1.68P/bt

(b) From Eq. (h) of Example 5.10,

$$Z=-1+\frac{3}{4}b(\frac{2}{bt})t\cdot\ln 2=0.0397$$

Points A and B. Apply Eqs.(5.71) with P=P/2 and M=-0.136Pb:

$$(\sigma_\theta)_A=\frac{P}{bt}+\frac{-0.136\,Pb}{3tb^2/8}\left[1+\frac{-b/4}{0.0397(b/2)}\right]$$

=5.21P/bt ◄

$$(\sigma_\theta)_B=\frac{P}{bt}+\frac{-0.136\,Pb}{3tb^2/8}\left[1+\frac{b/4}{0.0397(b)}\right]$$

=-1.65P/bt ◄

(c) $(\sigma_\theta)_{A,B}=(P/2A)\pm(Mc/I)$
Here c=b/4 I=$tb^3/96$
Thus,

$$(\sigma_\theta)_A=\frac{P}{bt}+\frac{0.136\,Pb\,(b/4)}{tb^3/96}$$

$$=\frac{P}{bt}+3.26\frac{P}{bt}=4.26P/bt$$ ◄

and

$$(\sigma_\theta)_B=\frac{P}{bt}-3.26\frac{P}{bt}=-2.26P/bt$$ ◄

Comparing the result, we observe the following differences:
At point A:
Elasticity vs. Winkler 0.2%
Elasticity vs. Elementary 18% ◄
At point B:
Elasticity vs. Winkler 1.8%
Elasticity vs. Elementary 34.5% ◄

5.34

From the condition of symmetry, the distribution of stress in any quadrant is known to be the same as the other.

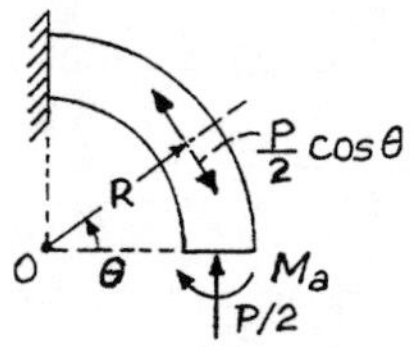

At any angle θ, the bending moment referring to this figure is expressed as follows

$$M_\theta=M_a-\frac{1}{2}PR(1-\cos\theta) \qquad (a)$$

The problem is statically indeterminate and value of M_a is found first. Note that shear component of the load will be omitted in our solution.

Applying Castigliano's theorem, we have:

$$\frac{\partial U}{\partial M_a}=4\int_0^{\pi/2}\frac{1}{EI}M_\theta\frac{\partial M_\theta}{\partial M_a}ds$$

$$0=\frac{4}{EI}\int_0^{\pi/2}[M_a-\frac{PR}{2}(1-\cos\theta)]Rd\theta$$

$$0=\frac{4}{EI}[\frac{\pi}{2}M_a-\frac{PR}{2}(\frac{\pi}{2}-1)]$$

Solving,

$$M_a=PR(\frac{1}{2}-\frac{1}{\pi})=0.182PR$$

Then, Eq. (a) becomes

$$M_\theta=0.182PR-\frac{1}{2}PR(1-\cos\theta) \qquad (b)$$ ◄

Therefore stress at any point of a section, using Eq. (5.71), is expressed in the form

$$\sigma_\theta=-\frac{P}{2}\frac{\cos\theta}{A}+\frac{M_\theta}{AR}\left[1+\frac{y}{Z(R+y)}\right]$$ ◄

Here, the moment M_θ is given by Eq. (b).

5.35

(a) Using Eq. (P5.34) at $\theta=\pi/4$ (Fig. P5.34):

$$M_\theta=0.182(1.5)P-\frac{1}{2}P(1.5)(1-\cos\frac{\pi}{4})=0.00533P \text{ N}\cdot\text{m}$$

We have

$$A=0.05(0.1)=0.005 \text{ m}^2$$

and

$$Z=-1+\frac{0.15}{0.1}\ln\frac{0.15+0.05}{0.15-0.05}=0.03972$$

At inner fiber, y=-0.05 m:

$$(\sigma_\theta)_{\pi/4}=\frac{-(P/2)\cos(\pi/4)}{0.005}+\frac{0.00533P}{0.005\times0.15}\times\left[1+\frac{-0.05}{0.03972(0.1)}\right]$$

$$=-152.87P \text{ Pa}$$ ◄

(b)

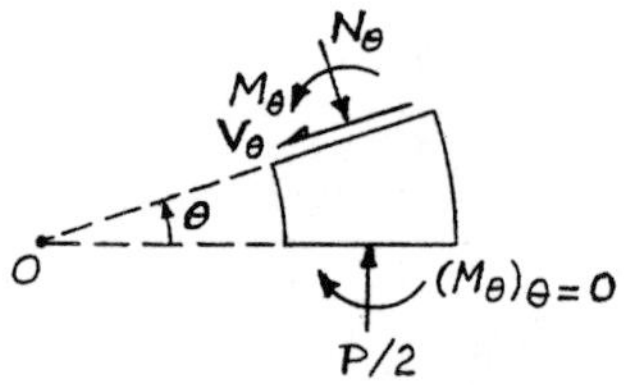

At $\theta=0°$:

$M_\theta=0.182PR$

$N_\theta=-P/2$

At any angle:

$$M_\theta=0.182PR-0.5PR(1-\cos\theta)=-0.318PR+0.5PR\cos\theta$$

$$\frac{\partial M_\theta}{\partial(P/2)}=-0.636R+R\cos\theta$$

$$N_\theta=-\frac{P}{2}\cos\theta, \quad \frac{\partial N_\theta}{\partial(P/2)}=-\cos\theta$$

$$V_\theta=-\frac{P}{2}\sin\theta, \quad \frac{\partial V_\theta}{\partial(P/2)}=-\sin\theta$$

Based on the symmetry of the ring, we have

$$\delta=2\int_0^{\pi/2}\left[\frac{N_\theta}{AE}\frac{\partial N_\theta}{\partial(P/2)}+\frac{M_\theta}{EI}\frac{\partial M_\theta}{\partial(P/2)}+\frac{\alpha V_\theta}{AG}\frac{\partial V_\theta}{\partial(P/2)}\right]Rd\theta$$

(CONT.)

5.35 CONT.

Substituting the values of M_θ, N_θ, and V_θ, this expression becomes

$$\delta_P=2\int_0^{\pi/2}\left[\frac{P}{AE}\cos^2\theta+\frac{PR^2}{2EI}(-0.636+\cos\theta)^2+\frac{\alpha}{AG}\frac{P}{2}\sin^2\theta\right]Rd\theta$$

$$=2\left[\frac{PR}{2AE}\left|\frac{\theta}{2}+\frac{1}{4}\sin2\theta\right|_0^{\pi/2}+\frac{PR^3}{2EI}\left|0.904\theta-1.27\sin\theta+\frac{\sin2\theta}{4}\right|_0^{\pi/2}+\frac{\alpha PR}{2AG}\left|\frac{\theta}{2}-\frac{1}{4}\sin2\theta\right|_0^{\pi/2}\right]$$

from which

$$\delta_P=2\left[\frac{PR}{2AE}\left(\frac{\pi}{4}\right)+\frac{PR^3}{2EI}(0.15)+\frac{\alpha}{AG}\frac{PR}{2}\left(\frac{\pi}{2}\right)\right]$$ ◄

For the given rectangular cross section:

$R=0.15$ m $\qquad A=0.005$ m^2

$\alpha=\frac{6}{5}=1.2 \qquad G=\frac{2}{5}E=0.4E$

$I=\frac{1}{12}(0.05)(0.1)^3=4.17(10^{-6})$ m^4

The deflection is therefore

$$\delta_P=2\left[\frac{P(0.15)\pi}{0.005E(8)}+\frac{P(0.15)^3(0.15)}{2E(4.17\times10^{-6})}+\frac{1.2}{0.4E(0.005)}\frac{P(0.15)\pi}{8}\right]$$

This results in

$$\delta_P=215.65P/E \text{ m}$$ ◄

6.1

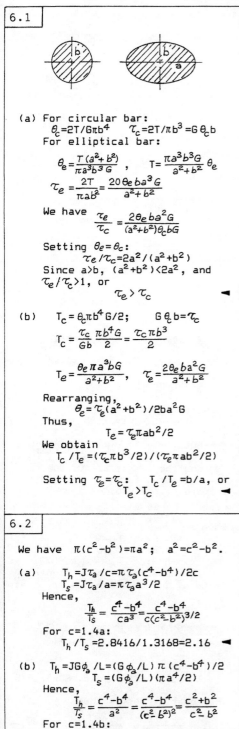

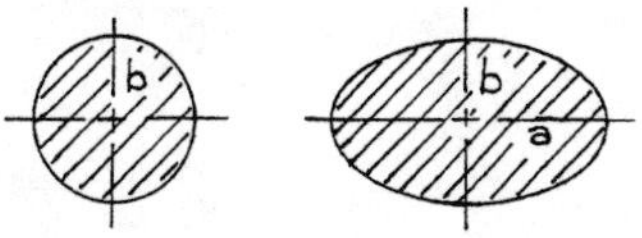

(a) For circular bar:

$\theta_c = 2T/G\pi b^4 \qquad \tau_c = 2T/\pi b^3 = G\theta_c b$

For elliptical bar:

$$\theta_e = \frac{T(a^2+b^2)}{\pi a^3 b^3 G}, \qquad T = \frac{\pi a^3 b^3 G}{a^2+b^2}\theta_e$$

$$\tau_e = \frac{2T}{\pi a b^2} = \frac{2\theta_e b a^3 G}{a^2+b^2}$$

We have $\dfrac{\tau_e}{\tau_c} = \dfrac{2\theta_e b a^2 G}{(a^2+b^2)\theta_c b G}$

Setting $\theta_e = \theta_c$:

$\tau_e/\tau_c = 2a^2/(a^2+b^2)$

Since $a>b$, $(a^2+b^2)<2a^2$, and $\tau_e/\tau_c>1$, or

$$\tau_e > \tau_c$$ ◄

(b) $T_c = \theta_c \pi b^4 G/2; \qquad G\theta_c b = \tau_c$

$$T_c = \frac{\tau_c}{Gb}\frac{\pi b^4 G}{2} = \frac{\tau_c \pi b^3}{2}$$

$$T_e = \frac{\theta_e \pi a^3 b G}{a^2+b^2}, \qquad \tau_e = \frac{2\theta_e b a^2 G}{a^2+b^2}$$

Rearranging,

$\theta_e = \tau_e(a^2+b^2)/2ba^2G$

Thus,

$T_e = \tau_e \pi a b^2/2$

We obtain

$T_c/T_e = (\tau_c \pi b^3/2)/(\tau_e \pi a b^2/2)$

Setting $\tau_e = \tau_c$: $T_c/T_e = b/a$, or

$T_e > T_c$ ◄

6.2

We have $\pi(c^2-b^2) = \pi a^2; \quad a^2 = c^2-b^2$.

(a) $T_h = J\tau_a/c = \pi\tau_a(c^4-b^4)/2c$

$T_s = J\tau_a/a = \pi\tau_a a^3/2$

Hence,

$$\frac{T_h}{T_s} = \frac{c^4-b^4}{ca^3} = \frac{c^4-b^4}{c(c^2-b^2)^{3/2}}$$

For $c=1.4a$:

$T_h/T_s = 2.8416/1.3168 = 2.16$ ◄

(b) $T_h = JG\phi_a/L = (G\phi_a/L)\pi(c^4-b^4)/2$

$T_s = (G\phi_a/L)(\pi a^4/2)$

Hence,

$$\frac{T_h}{T_s} = \frac{c^4-b^4}{a^2} = \frac{c^4-b^4}{(c^2-b^2)^2} = \frac{c^2+b^2}{c^2-b^2}$$

For $c=1.4b$:

$T_h/T_s = 2.96/0.96 = 3.08$ ◄

6.3

Use Eqs. (6.4):

$$T_A = \frac{T}{1+(aJ_b/bJ_a)} = \frac{T}{1+\dfrac{0.4(15)^4(\pi/32)}{0.2(20)^4(\pi/32)}}$$

$= T/(1+0.6328) = 0.6124T$

$T_B = T - T_A = 0.3876T$

Based on shear in segment AC:

$$\tau_a = \frac{16T_A}{\pi d_a^3} = \frac{16(0.6124)T}{\pi(0.02)^3} = 150(10^6)$$

or

$T = 384.7$ N·m

Based on shear in segment CB:

$$\frac{16T_B}{\pi d_b^3} = \frac{16(0.3876)T}{\pi(0.015)^3} = 150(10^6)$$

or

$T = 256.5 \text{ N·m} = T_{all}$ ◄

6.4

Use Eqs. (6.4):

$$T_A = \frac{T}{1+\dfrac{0.8(15)^4(\pi/32)}{0.5(25)^4(\pi/32)}} = \frac{T}{1+0.2074}$$

$= 0.8283T$

$T_B = T - T_A = 0.1717T$

Based on shear in segment AC:

$$\frac{16T_A}{\pi d_a^3} = \frac{16(0.8283)T}{\pi(0.025)^3} = 70(10^6)$$

or

$T = 259.3 \text{ N·m} = T_{all}$ ◄

Based on shear in segment CB:

$$\frac{16T_B}{\pi d_b^3} = \frac{16(0.1717)T}{\pi(0.015)^3} = 70(10^6)$$

or

$T = 270.2$ N·m

6.5

$$\frac{\partial^2\Phi}{\partial x^2} = k[2(a^2b-a^2)+2y^2(b^2+1)-12bx^2]$$

$$\frac{\partial^2\Phi}{\partial y^2} = k[2(a^2b-a^2)+2x^2((b^2+1)-12by^2]$$

Substituting these in Eq. (6.9):

$$k = \frac{G\theta}{2a^2(b-1)+(b^2-6b+1)(x^2+y^2)}$$

In order k be a constant

$b^2-6b+1=0$

Thus,

$$k = \frac{G\theta}{2a^2(b-1)}$$ ◄

6.6

$u=-\theta z(y-b)$, $\vartheta=\theta z(x-z)$, $w=w(x,y)$
$\varepsilon_x=\varepsilon_y=\varepsilon_z=0$, $\gamma_{xy}=(\partial v/\partial x)+(\partial u/\partial y)=0$

Thus

$$\gamma_{xz}=\frac{\partial w}{\partial x}-\theta(y-b),\qquad \gamma_{yz}=\frac{\partial w}{\partial y}+\theta(x-a)$$

$$\tau_{xz}=G\left[\frac{\partial w}{\partial x}-\theta(y-b)\right]=\frac{\partial\Phi}{\partial y} \quad ◄$$

$$\tau_{yz}=G\left[\frac{\partial w}{\partial y}+\theta(x-a)\right]=-\frac{\partial\Phi}{\partial x}$$

Substituting these in Eqs. (6.6), (6.7), and (6.11), we obtain

$$\frac{\partial\tau_{xz}}{\partial y}-\frac{\partial\tau_{yz}}{\partial x}=-2G\theta,\ \frac{\partial^2\Phi}{\partial x^2}+\frac{\partial^2\Phi}{\partial y^2}=-2G\theta \quad ◄$$

and

$$T=\iint[(x-a)\tau_{yz}-(y-b)\tau_{xz}]dxdy$$
$$=\left[-\iint(x-a)\frac{\partial\Phi}{\partial x}-\iint(y-b)\frac{\partial\Phi}{\partial y}\right]dxdy$$
$$=2\iint\Phi\, dxdy \quad ◄$$

We observe that the characteristic equations remain unchanged.

6.7

$$T=\iint(x\,\tau_{yz}-y\,\tau_{xz})dxdy$$
$$=-\iint\left(x\frac{\partial\Phi}{\partial x}-y\frac{\partial\Phi}{\partial y}\right)dxdy$$

Here

$$-\int dy\int x\frac{\partial\Phi}{\partial x}dx$$
$$=-\int x d\Phi\Big|_{x_1}^{x_2}dy-\left[-\iint\Phi dxdy\right]$$
$$=-c\int(x_2-x_1)dy+\iint\Phi dxdy$$
$$=-c\iint dxdy+\iint\Phi dxdy$$

Note that $(x_2-x_1)dy$ is area and equals $\iint dxdy$. Similarly,

$$-\int dx\int y\frac{\partial\Phi}{\partial y}dy=-c\int\int dydx+\iint\Phi\, dxdy$$

Thus,

$$T=2\iint(\Phi-c)dxdy \quad ◄$$

6.8

$$T_\theta=T\cos^2\theta$$

Hence,

$$\theta=\int\frac{T_\theta dz}{GJ}=\frac{1}{GJ}\int T\cos^2\theta\, a d\theta$$

This gives, at sections A and B:

$$\theta_A=\frac{Ta}{GJ}\int_0^{\pi/2}\left(\frac{1}{2}+\frac{\cos 2\theta}{2}\right)d\theta=\frac{Ta}{2r^4G} \quad ◄$$

$$\theta_B=\frac{Ta}{GJ}\int_0^{\pi}\cos^2\theta\, d\theta=\frac{Ta}{r^4G}$$

6.9

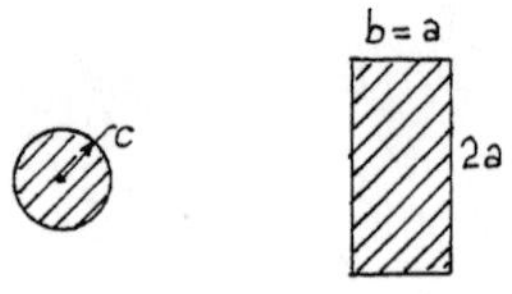

From Table 6.2: $\alpha=0.246$, $\beta=0.229$

$$\tau_r=T/\alpha ab^2=T/0.492a^3$$
$$\tau_c=Tc/J=2T/\pi c^3$$

Thus,

$$T/0.492a^3=2T/\pi c^3;\quad c=0.684a$$

Similarly,

$$\theta_r=\frac{T}{0.229(2a)a^3G}=\frac{T}{0.458a^4G}$$
$$\theta_c=T/JG=2T/\pi c^4G$$

Then,

$$\frac{T}{0.458a^4G}=\frac{2T}{\pi c^4G}$$

gives

$$c=0.736a=c_{all} \quad ◄$$

6.10

$$\frac{\partial^2\Phi}{\partial y^2}=k\left[\left(x+\frac{h}{3}\right)+\left(x+\sqrt{3}\,y-\frac{2}{3}h\right)+\left(x+\frac{h}{3}\right)+\left(x-\sqrt{3}\,y-\frac{2}{3}h\right)+2\left(x-\frac{2}{3}h\right)\right]$$

$$\frac{\partial^2\Phi}{\partial y^2}=k\left[-3\left(x+\frac{h}{3}\right)-3\left(x+\frac{h}{3}\right)\right]$$

Substituting this in Eq. (6.9),

$$-4kh=-2G\theta\ ;\ k=G\theta/2h$$

Thus,

$$\Phi=-G\theta\left[\frac{1}{2}(x^2+y^2)-\frac{1}{2h}(x^3-3xy^2)-\frac{2}{27}h^2\right]$$

Along the x axis $\tau_{xz}=0$, due to the symmetry. Equation (6.8) is therefore, for $y=0$:

$$\tau_{yz}=-\frac{\partial\Phi}{\partial x}=\frac{3G\theta}{2h}\left(\frac{2hx}{3}-x^2\right)$$

When

$x=0$: $\tau_{yz}=0$
$x=2h/3$: $\tau_{yz}=0$
$x=-h/3$: $\tau_{yz}=\tau_{max}=G\theta h/2$ ◄

Next, substitute the preceding value of Φ into Eq.(6.11) to obtain

$$T=2\iint\Phi dxdy$$
$$=-4G\theta\int_0^{h/\sqrt{3}}\int_0^{\sqrt{3}y+\frac{2}{3}h}\left[\frac{1}{2}(x^2+y^2)-\frac{1}{2h}(x^3-3xy^2)-\frac{2}{27}h^2\right]dxdy=G\theta h^4/15\sqrt{3}$$

The shear stress is thus

$$\tau_{max}=\frac{20T}{(2h/\sqrt{3})^3}=\frac{15\sqrt{3}\,T}{2h^3} \quad ◄$$

6.11

For a circular bar:

$$T = G\theta J = C_c \theta$$

where

$$C_c = \pi r^4 G/2 \qquad \text{(a)}$$

For an elliptical bar (from Example 6.2):

$$H = -2T(a^2+b^2)/\pi a^3 b^3 = -2G\theta$$

$$T = \frac{\pi a^3 b^3}{a^2+b^2} G\theta = C_e \theta$$

or

$$C_e = \frac{\pi a^3 b^3}{a^2+b^2} G \qquad \text{(b)}$$

For an equilateral bar (from Prob. 6.10):

$$T = G\theta h^4/15\sqrt{3} = \theta C_t$$

or

$$C_t = h^4 G/15\sqrt{3} \qquad \text{(c)}$$

Bars have equal areas:

$$A_c = \pi r^2, \quad A_e = \pi ab, \quad A_t = h^2/\sqrt{3}$$

Setting $A_c = A_e = A_t$:

$$r^4 = a^2 b^2 = h^4/3\pi^2$$

Then, Eqs. (a), (b), and (c) give

$$\frac{C_e}{C_c} = \frac{2a^3 b^3}{r^4(a^2+b^2)} = \frac{2ab}{a^2+b^2} \quad ◄$$

and

$$\frac{C_t}{C_c} = \frac{h^4 G}{15\sqrt{3}} \frac{2}{\pi r^4 G} = \frac{2\pi\sqrt{3}}{15} \quad ◄$$

6.12

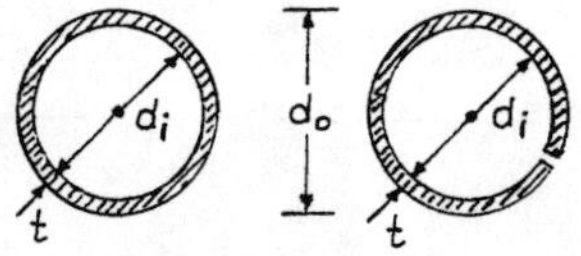

For the seamless tube, from $\theta_1 = T/GJ$:

$$\theta_1 = \frac{32T}{\pi d_o^4 (1 - d_i^4/d_o^4) G}$$

For the split tube, referring to Eq. (6.17):

$$\theta_2 = \frac{3T}{G} \frac{1}{\pi \left(\frac{d_o+d_i}{2}\right)\left(\frac{d_o-d_i}{2}\right)^3}$$

We have

$$\frac{\theta_1}{\theta_2} = \frac{2(d_o-d_i)^2}{3(d_o^2+d_i^2)} \qquad \text{(a)} \quad ◄$$

For <u>very thin</u> tubes $d_o^2 + d_i^2 \approx 2d_o^2$, and Eq. (a) becomes

$$\frac{\theta_1}{\theta_2} = \frac{4}{3}\left(\frac{t}{d_o}\right)^2 \quad ◄$$

6.13

Apply Eqs. (6.17) and (6.19):

$$\tau_{max} = \frac{3T}{bt^2} = \frac{3(80)}{0.125(0.005)^2} = 76.8 \text{ MPa} \quad ◄$$

$$\theta = \frac{3T}{bt^3 G} = \frac{3(80)}{0.125(0.005)^3(80\times10^9)} \quad ◄$$

$$= 0.192 \text{ rad/m}$$

6.14

Equation (6.20) yields

$$J_e = \sum \tfrac{1}{3} bt^3 = \tfrac{1}{3}(100)(10)^3 + \tfrac{1}{3}(115)(4)^3$$

$$= 3.5787(10^4) \text{ mm}^4 \quad ◄$$

Maximum shear stress occurs on the lower leg:

$$\tau_{max} = \frac{Tt_1}{J_e} = \frac{500(0.1)}{3.5787(10^{-8})} = 139.7 \text{ MPa} \quad ◄$$

Angle of twist per unit length is

$$\theta = T/J_e G$$

$$= \frac{500}{3.5787(10^{-8})(200\times10^9)}$$

$$= 69.86\ (10^{-3}) \text{ rad/m}$$

$$= 4.00 \text{ deg per meter} \quad ◄$$

6.15

Refer to Table 6.2.

(a)

$$T = \frac{a_2^3 \tau_{max}}{20} = \frac{(0.045)^3(50\times10^6)}{20} = 228 \text{ N·m}$$

$$T = a_2^4 G\theta_{all}/46.2$$

$$= \frac{(0.045)^4(80\times10^9)(1.5\pi/180)}{46.2}$$

$$= 186 \text{ N·m} = T_{all} \quad ◄$$

(b)

$$\phi = \frac{46.2\, T_{max}}{G}\left[\frac{L_1}{a_1^4} + \frac{L_2}{a_2^4}\right]$$

$$= \frac{46.2(186)}{80\times10^9}\left[\frac{2.5}{(0.06)^4} + \frac{1.5}{(0.045)^4}\right]$$

$$= 1.0742(10^{-7})[192{,}901.235 + 365{,}797.897]$$

$$= 0.06 \text{ rad} = 3.44° \quad ◄$$

6.16

Referring to Fig. P6.16, an expression for t is written as

$$t=t_0(1-y/b)$$

Substitute this into the given stress function to obtain,

$$\Phi = G\theta\left[\frac{t_0^2}{4}\left(1-\frac{y}{b}\right)^2 - x^2\right]$$

We have

$$T=2\iint \Phi\, dx\,dy$$

$$=4\int_0^{t_0/2}\int_0^{b(1-\frac{2x}{t_0})}\left[\frac{t_0^2}{4}\left(1-\frac{y}{b}\right)^2 - x^2\right]dx\,dy$$

$$=4G\theta\int_0^{t_0/2}\left\{\frac{t_0^2}{4}\left[b\left(1-\frac{2x}{t_0}\right) - b\left(1-\frac{2x}{t_0}\right)^2 + \frac{b}{3}\left(1-\frac{2x}{t_0}\right)^3\right] - bx^3\left(1-\frac{2x}{t_0}\right)\right\}dx$$

Let $u=(1-2x)/t_0$, $x=(1-u)t_0/2$

$du=-2dx/t_0$, $dx=-t_0\,du/2$

Then, the preceding expression for the torque becomes

$$T=4G\frac{t_0^3 b}{8}\int_0^1\left(u^2-\frac{2}{3}u^3\right)du$$

Integrating,

$$T=G\theta t_0^3 b/12$$ ◄

6.17

(a) From Table 6.2: $T=(2\pi r^3 tG)\theta=C\theta$

or

$C=2\pi r^3 tG=2\pi(0.02375)^3(0.0025)G$

$=2.1(10^{-7})G$

$$\tau_{max}=\frac{T}{2\pi r^2 t}=\frac{T}{2\pi(0.02375)^2(0.0025)}$$

$=112,860T$ ◄

(b) Equation (6.16):

$C=bt^3G/3$

$=[2\pi(0.02375)(0.0025)^3 G]/3$

$=8(10^{-10})G$

Equation (6.18):

$$\tau_{max}=\frac{3T}{bt^2}=\frac{3T}{2\pi(0.02375)(0.0025)^2}$$

$=3,252,032T$ ◄

(c) From Table 6.2: for a=b, $t=t_1$:

$T=a^3tG\theta=1.0717(10^{-4})0.0025G\theta$

$=2.7(10^{-7})G\theta=C\theta$

$$\tau_{max}=\frac{T}{2a^2t}=\frac{T}{2(0.00225625)(0.0025)}$$

$=88,643T$ ◄

6.18

Referring to Table 6.2:

$$\tau_A=20T/a^3$$

or

$$420(2)/3=20T/(0.05)^3$$

Solving,

$$T=1.75\ \text{kN}\cdot\text{m}$$ ◄

Also,

$$\theta=46.2T/a^4G$$

$$=\frac{46.2\,(1.75\times10^3)}{(6.25\times10^{-6})(80\times10^9)}$$

$=0.1617$ rad/m ◄

6.19

For a thin-walled tube, letting

$$r_o\approx r_i\approx r_{avg.}=r$$

$$J=\pi(r_o^4-r_i^4)/2$$

$$=\frac{\pi}{2}(r_o^2+r_i^2)(r_o+r_i)(r_o-r_i)=2\pi rt$$

Equation (6.2):

$$\tau=\frac{T\rho}{J}=\frac{T\rho}{2\pi r^3t}=\frac{T\rho}{2Art}$$

Since $r=\rho$, we obtain Eq. (6.22): $\tau=T/2At$. ◄

6.20

For a regular hexagon, we can write

$$A=\frac{3\sqrt{3}}{2}a^2 \qquad ds=6a$$

Equation (6.22), substitiuting the value of A, results in the shear stress

$$\tau=\frac{T}{2At}=\frac{T\sqrt{3}}{9a^2t}$$ ◄

Angle of twist per unit length, using Eq. (6.23):

$$\theta=\frac{\tau(6a)}{2GA}=\frac{3Ta}{2A^2Gt}$$

or

$$\theta=2T/9Ga^3t$$ ◄

6.21

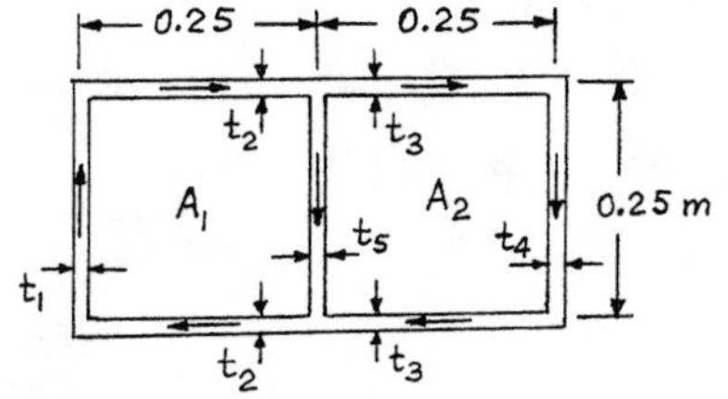

Given:
$t_1=0.012$ m, $t_2=t_3=0.006$ m,
$t_4=t_5=0.01$ m, G=28 GPa
T=56.5 kN·m, $A_1=A_2=A=0.0625$ m^2
$s_1=s_2=s_3=s_4=s_5=0.25$ m.

We write

$$T=2A_1h_1+2A_2h_2=2A_1t_1\tau_1+2A_2t_3\tau_3$$

or

$$2\tau_1+\tau_3=75.3(10^6) \qquad (1)$$

The shear flow yields

$$\tau_1t_1=\tau_2t_2 \qquad (2)$$
$$\tau_2t_2=\tau_3t_3+\tau_5t_5 \qquad (3)$$
$$\tau_3t_3=\tau_4t_4 \qquad (4)$$

Also,

$$\tau_1s_1+2\tau_2s_2+\tau_5s_5=2G\theta A_1 \qquad (5)$$
$$-\tau_5s_5+2\tau_3s_3+\tau_4s_4=2G\theta A_2 \qquad (6)$$

Simultaneous solution of Eqs. (1) through (6), after substitutiton of the given numerical values, yields:

$\tau_1=19.1$ MPa $\quad \tau_2=38.2$ MPa
$\tau_3=37.2$ MPa $\quad \tau_4=22.3$ MPa ◄
$\tau_5=0.62$ MPa

and

$$\theta=6.86(10^{-3}) \text{ rad/m}$$ ◄

6.22

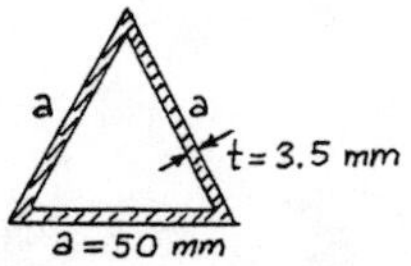

We have $A=(0.05)(0.05\sin 60°)/2 = 1.0825(10^{-3})$ m^2. Then,

$$h=\frac{T}{2A}=\frac{40}{2(0.0010825)}=18{,}475.209 \text{ N/m}$$

$$\tau=\frac{h}{t}=\frac{18{,}475.209}{0.0035}=5.279 \text{ MPa}$$ ◄

$$\theta=\frac{\tau}{2GA}\int ds=\frac{5.279(10^6)}{2(28\times10^9)(0.0010825)}\times[0.05+0.05+0.05]$$

$$=13.063(10^{-3}) \text{ rad/m}$$ ◄

6.23

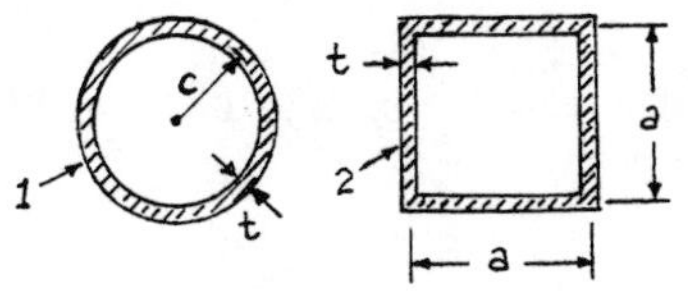

For circular tube:
Area enclosed by c, $A_{m1}=\pi c^2$
Area of the section, $A_1=2\pi ct$
Polar moment of inertia, $J_1=2\pi c^3t$

For square tube:
Enclosed area by a, $A_{m2}=a^2$
Area of the section, $A_2=4at$
Polar moment of inertia (from Table 6.2 with $t=t_1$ and $a=b$):

$$J_2=2tt_1a^2b^2/(at+bt_1)=a^3t$$

Then, $A_1=A_2$ gives: $a=\pi c/2$ and hence, from Eq. (6.22) we obtain

$$\tau_1/\tau_2=A_{m2}/A_{m1}=a^2/\pi c^2=0.785$$ ◄

Similarly, from $\phi=T/GJ$:

$$\phi_1/\phi_2=J_2/J_1=a^3/2\pi c^3=0.617$$ ◄

6.24

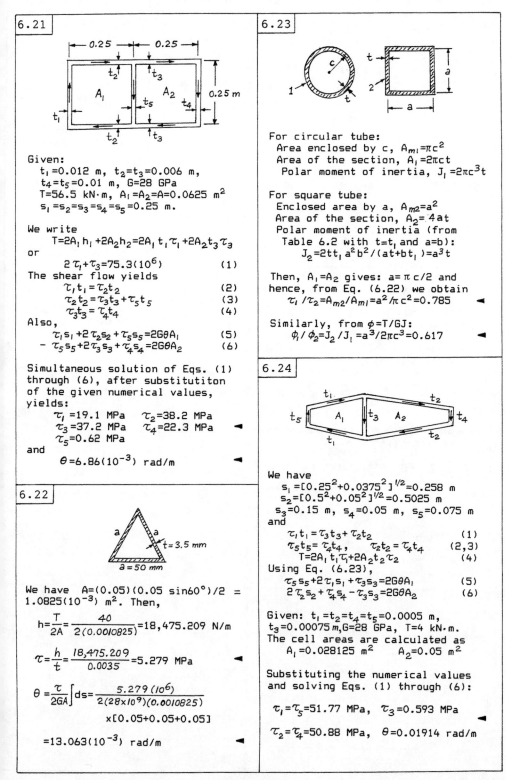

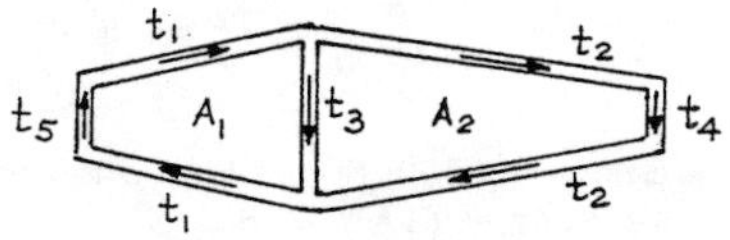

We have

$s_1=[0.25^2+0.0375^2]^{1/2}=0.258$ m
$s_2=[0.5^2+0.05^2]^{1/2}=0.5025$ m
$s_3=0.15$ m, $s_4=0.05$ m, $s_5=0.075$ m

and

$$\tau_1t_1=\tau_3t_3+\tau_2t_2 \qquad (1)$$
$$\tau_5t_5=\tau_4t_4, \quad \tau_2t_2=\tau_4t_4 \qquad (2,3)$$
$$T=2A_1t_1\tau_1+2A_2t_2\tau_2 \qquad (4)$$

Using Eq. (6.23),

$$\tau_5s_5+2\tau_1s_1+\tau_3s_3=2G\theta A_1 \qquad (5)$$
$$2\tau_2s_2+\tau_4s_4-\tau_3s_3=2G\theta A_2 \qquad (6)$$

Given: $t_1=t_2=t_4=t_5=0.0005$ m, $t_3=0.00075$ m, G=28 GPa, T=4 kN·m.
The cell areas are calculated as
$A_1=0.028125$ m^2 $\quad A_2=0.05$ m^2

Substituting the numerical values and solving Eqs. (1) through (6):

$\tau_1=\tau_5=51.77$ MPa, $\tau_3=0.593$ MPa ◄

$\tau_2=\tau_4=50.88$ MPa, $\theta=0.01914$ rad/m

6.25

(a) We have $\delta_s = \delta_c$ and Eq. (6.37) becomes

$$\frac{P_s R_s^3}{G_s} = \frac{P_c R_c^3}{G_c}$$

or

$$\frac{P_s (0.062)^3}{79(10^9)} = \frac{P_c (0.5)^3}{41(10^9)}$$

Solving,

$$P_s = 1.012 P_c$$

Assume that copper controls

$$\tau_c = \frac{16 P_c R_c}{\pi d^3}$$

$$300(10^6) = \frac{16 P_c (0.05)}{\pi (0.01)^3}$$

from which

$$P_c = 1175.5 \text{ N}$$

Then,

$$P_s = 1189.6 \text{ N}$$

Check the assumption:

$$\tau_s = \frac{16(1189.6)\ 0.062)}{\pi (0.01)^3} = 375.82 \text{ MPa}$$

Since $\tau_s < 500$ MPa, the assumption is correct.

Thus, the total force is

$$P_t = P_c + P_s = 2365.1 \text{ N}$$ ◄

(b) For the condition specified,

$$P_s k_s = P_c k_c$$

from which

$$\frac{k_c}{k_s} = \frac{P_s}{P_c} = 1.012$$ ◄

7.1

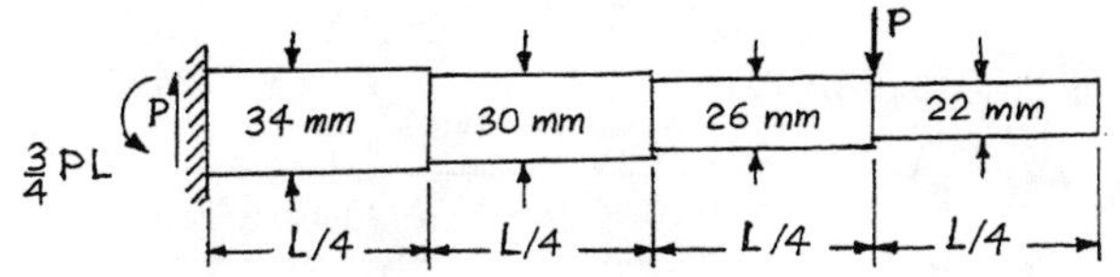

Fig. (a)

Equivalent beam is in Fig. (a). Referring to this figure d_n, EI_n, and M_n are found for each segment. Then, Eqs. (7.1) and (7.2) are applied together with the boundary conditions $\theta(0)=0$ and $\upsilon(0)=0$. Results are presented in Table P7.1.

Table P7.1

Units	n	1	2	3	4
m	d_n	0.034	0.030	0.026	0.022
$N\cdot m^2$	$(EI_n)10^{10}$	655E	397E	224E	115E
N·m	M_n	0.625PL	0.375PL	0.125PL	0
rad	$(\theta_n)10^{-4}$	$238PL^2/E$	$475PL^2/E$	$614PL^2/E$	$614PL^2/E$
m	$(\upsilon_n)10^{-3}$	0	$595PL^3/E$	$1780PL^3/E$	$3320PL^3/E$

7.2

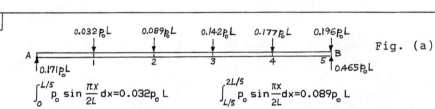

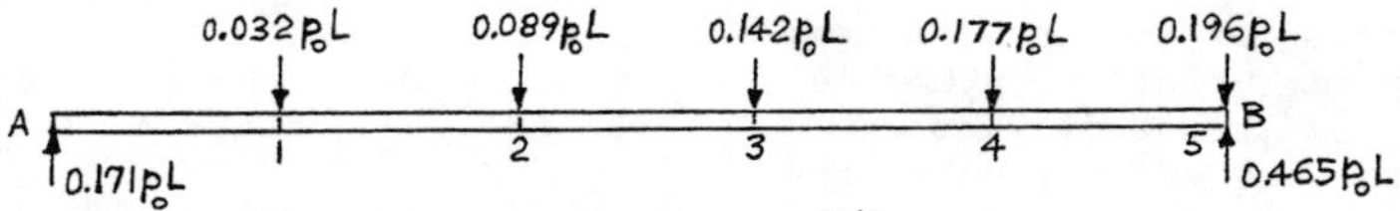

Fig. (a)

$$\int_0^{L/5} p_o \sin\frac{\pi x}{2L}dx=0.032p_o L \qquad \int_{L/5}^{2L/5} p_o \sin\frac{\pi x}{2L}dx=0.089p_o L$$

Similarly, equivalent loading at the remaining points are found (Fig.a).
Moments $(M_n/p_o L^2)$:

$M_1=0.171L/10=-0.017$, $M_2=-0.171(3/10)+0.032/10=-0.048,\ldots,$.

Following a procedure similar to that used in Example 7.1 we determine the quantities listed in Table P7.2. Sample calculations are as follows:

$(\theta_n-\theta_o)/(p_o L^3/EI)$:

$M_1\Delta x=-0.0034$ (1), $M_1\Delta x+M_2\Delta x=-(0.003+0.01)=-0.013$ (2)

$(\upsilon_n-n\,\Delta x\cdot\theta_o)/(p_o L^4/EI)$:

0 (1), $M_1\Delta x\cdot\Delta x=-0.0034/5=-0.0007$ (2)

Since $\upsilon_5=0$: $0-5(L/5)\theta_o=-0.0163p_o L^4/EI$ or $\theta_o=0.0163p_o L^3/EI$

$\theta_n/(p_o L^3/EI)$:

0.0129 (1), -0.0033 (2), 0.0099 (3), -0.0227 (4), -0.028 (5)

$\upsilon_n/(p_o L^4/EI)$:

0.0163/5=-0.0033, (1), 0.0033+(0.0129/5)=0.0059 (2),...,.

Table P7.2

Units	n	1	2	3	4	5
N·m	M_n/A	-0.17	-0.48	-0.67	-0.063	-0.0267
rad	$(\theta_n-\theta_o)/B$	-0.0034	-0.013	-0.026	-0.039	-0.044
m	$[\upsilon_n-n(\Delta x)\theta_o]/C$	0	-0.0007	-0.0033	-0.0085	-0.0163
rad	θ_n/B	0.0129	0.0033	-0.0097	-0.0227	-0.028
m	υ_n/C	0.0033	0.0059	0.0065	0.0045	0

Note: The constants are $A=p_o L^2$, $B=p_o L^3/EI$, and $C=p_o L^4/EI$.

7.3

Refer to Fig. 7.4 and Eq. (7.12):

$$\frac{\partial^4 w}{\partial x^4} = \frac{1}{h^4}(w_9 - 4w_1 + 6w_0 - 4w_3 + w_{11})$$

$$\frac{\partial^4 w}{\partial y^4} = \frac{1}{h^4}(w_{10} - 4w_2 + 6w_0 - 4w_4 + w_{12})$$

$$\frac{\partial^4 w}{\partial x^2 \partial y^2} = \frac{1}{h^4}[w_5 - w_6 + w_7 + w_8 - 2(w_1 - w_2 - w_3 - w_4 + 2w_0)]$$

Substitiuting these into $\nabla^4 w$:

$$h^4 \nabla^4 w = 20w_0 - 8(w_1 + w_2 + w_3 + w_4) - 2(w_5 + w_6 + w_7 + w_8) + w_9 + w_{10} + w_{11} + w_{12}$$

7.4

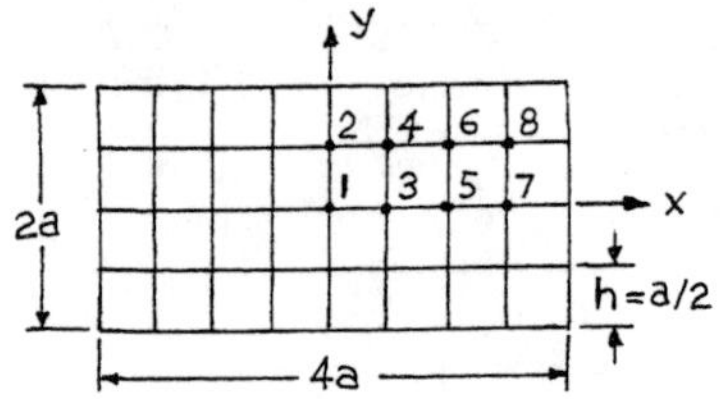

Only a quarter of the section need be considered. Using the symmetry the nodal points are labeled as shown in the preceding figure, and $\Phi = 0$ at the bondary. The finite difference equations for the nodes 1 through 8 are, respectively:

$$-4\Phi_1 + 2\Phi_2 + 2\Phi_3 = -2G\theta h^2 \quad (1)$$
$$\Phi_1 - 4\Phi_2 + 2\Phi_4 = -2G\theta h^2 \quad (2)$$
$$\Phi_1 - 4\Phi_3 + 2\Phi_4 + \Phi_5 = -2G\theta h^2 \quad (3)$$
$$\Phi_2 + \Phi_3 - 4\Phi_4 + \Phi_6 = -2G\theta h^2 \quad (4)$$
$$\Phi_3 - 4\Phi_5 + 2\Phi_6 + \Phi_7 = -2G\theta h^2 \quad (5)$$
$$\Phi_4 + \Phi_5 - 4\Phi_6 + \Phi_8 = -2G\theta h^2 \quad (6)$$
$$\Phi_5 - 4\Phi_7 + 2\Phi_8 = -2G\theta h^2 \quad (7)$$
$$\Phi_6 + \Phi_7 - 4\Phi_8 = -2G\theta h^2 \quad (8)$$

Solving,

$\Phi_1 = 3.587G\theta h^2$ $\Phi_2 = 2.708G\theta h^2$
$\Phi_3 = 3.467G\theta h^2$ $\Phi_4 = 2.622G\theta h^2$
$\Phi_5 = 3.037G\theta h^2$ $\Phi_6 = 2.313G\theta h^2$
$\Phi_7 = 2.055G\theta h^2$ $\Phi_8 = 1.592G\theta h^2$

Tables of differences are in P7.4a and P7.4b.

Table P7.4a
(when y=b, $\Phi=\Phi_1$; y=h, $\Phi=\Phi_2$, etc.)

y	$\Phi/G\theta h^2$	$\Delta/G\theta h^2$	$\Delta^2/G\theta h^2$
0	3.587	-0.380	-1.828
h	2.708	-2.708	
2h	0		

(CONT.)

7.4 CONT.

Table P7.4b
(when x=0, $\Phi=\Phi_1$; x=h, $\Phi=\Phi_3$, etc.)

x	$\Phi/G\theta h^2$	$\Delta/G\theta h^2$	$\Delta^2/G\theta h^2$	$\Delta^3/G\theta h^2$	$\Delta^4/G\theta h^2$
0	3.587	-0.120	-0.310	-0.242	-0.28
h	3.467	-0.430	-0.552	-0.522	
2h	3.037	-0.982	-1.074		
3h	2.055	-2.055			
4h	0				

We have

$$\frac{\partial \Phi}{\partial x} = \frac{\Delta \Phi_0}{h} + \frac{\Delta^2 \Phi_0}{2h^2}[2x-h] + \frac{\Delta^3 \Phi_0}{6h^3}[3x^2 - 6xh + 2h^2] + \frac{\Delta^4 \Phi_0}{24h^4}[4x^3 - 18x^2 h + 22xh^2 - 6h^3]$$

$$\left(\frac{\partial \Phi}{\partial x}\right)_{x=4h} = \frac{\Delta \Phi_0}{h} + \frac{\Delta^2 \Phi_0}{2h^2}(7h) + \frac{\Delta^3 \Phi_0}{6h^3}(26h^2) + \frac{\Delta^4 \Phi_0}{24h^4}(50h^3)$$

Using Table P7.4b, we obtain $(\partial\Phi/\partial x)_{x=4h} = -1.418G\theta a$. That is

$$(\tau)_{x=\pm 2a} = -1.418G\theta a$$ ◄

Similarly, using Table P7.4a:

$$\left(\frac{\partial \Phi}{\partial y}\right)_{y=2h} = \frac{\Delta \Phi_0}{h} + \frac{\Delta^2 \Phi_0}{2h^2}(2y-h) = -1.811G\theta a$$

$$(\tau)_{y=\pm 2h} = -1.811G\theta a$$ ◄

Hence, the error using this method is [(186-1.811)/1.86]100=2.63%

7.5

Equation (7.23) is applied at points b, c, d, respectively:

$$2\Phi_c + 1.178\Phi_g - 4.857\Phi_b = -2G\theta h^2$$
$$\Phi_b + 1.178\Phi_f + \Phi_d - 4.587\Phi_c = -2G\theta h^2$$
$$1.178\Phi_c + 1.178\Phi_e - 5.714\Phi_d = -2G\theta h^2$$

Similary, Eq. (7.18) is applied at points e, f, an g, respectively:

$$2\Phi_d + \Phi_f - 4\Phi_e = -2G\theta h^2$$
$$2\Phi_c + \Phi_g + \Phi_e - 4\Phi_f = -2G\theta h^2$$
$$2\Phi_b + 2\Phi_f - 4\Phi_g = -2G\theta h^2$$

In matrix form these equations are written as:

(CONT.)

7.5 CONT.

$$\begin{bmatrix} 2 & 0 & 0 & 0 & 2 & -4 \\ 0 & 2 & 0 & 1 & -4 & 1 \\ 0 & 0 & 2 & -4 & 1 & 0 \\ 0 & 1.178 & -5.714 & 1.178 & 0 & 0 \\ 1 & -4.857 & 1 & 0 & 1.178 & 0 \\ -4.857 & 2 & 0 & 0 & 0 & 1.178 \end{bmatrix} \begin{Bmatrix} \Phi_b \\ \Phi_c \\ \Phi_d \\ \Phi_e \\ \Phi_f \\ \Phi_g \end{Bmatrix}$$

$$=-2G\theta h^2\{1,1,1,1,1,1\}$$

Solution is

$$\Phi_b=1.612G\theta h^2 \qquad \Phi_c=1.487G\theta h^2$$
$$\Phi_d=0.975G\theta h^2 \qquad \Phi_e=1.547G\theta h^2$$
$$\Phi_f=2.236G\theta h^2 \qquad \Phi_g=2.424G\theta h^2$$

The finite differences are then

$$\Delta=\Phi_e-\Phi_B^{0}=1.547G\theta h^2$$

$$\Delta^2=\Phi_f-2\Phi_e+\Phi_B^{0}=-0.858G\theta h^2$$

$$\Delta^3=\Phi_g-3\Phi_f+3\Phi_e-\Phi_B^{0}=0.357G\theta h^2$$

$$\Delta^4=\Phi_f-4\Phi_g+6\Phi_f-4\Phi_B^{0}+\Phi_B^{0}=-0.232G\theta h^2$$

$$\Delta^5=\Phi_e-5\Phi_f+10\Phi_g-10\Phi_f+5\Phi_e-\Phi_B^{0}$$
$$=-0.018G\theta h^2$$

$$\Delta^6=\Phi_B^{0}-6\Phi_e+15\Phi_f-20\Phi_g+15\Phi_f-6\Phi_e+\Phi_B^{0}$$
$$=0.036G\theta h^2$$

Hence,

$$\tau_B=\left(\frac{\partial\Phi}{\partial x}\right)_B=\frac{1}{h}\left(\Delta-\frac{\Delta^2}{2}+\frac{\Delta^3}{3}-\frac{\Delta^4}{4}+\frac{\Delta^5}{5}-\frac{\Delta^6}{6}\right)G\theta h^2$$
$$=\left(1.547+\frac{0.858}{2}+\frac{0.357}{3}+\frac{0.232}{4}-\frac{0.018}{5}-\frac{0.036}{6}\right)G\theta h$$

or

$$\tau_B=2.144G\theta h=0.0107G\theta$$ ◄

7.6

Applying Eq. (7.18) at points b, c, e, f, and g, we obtain

$$\Phi_g+2\Phi_c-4\Phi_b=-2G\theta h^2$$
$$\Phi_b+\Phi_d+\Phi_f-4\Phi_c=-2G\theta h^2$$
$$2\Phi_d+\Phi_f-4\Phi_e=-2G\theta h^2$$
$$\Phi_g+\Phi_e+2\Phi_c-4\Phi_f=-2G\theta h^2$$
$$2\Phi_f+2\Phi_b-4\Phi_g=-2G\theta h^2$$

At point d, we apply Eq. (7.23):

$$1.308\Phi_c+\Phi_e-5.77\Phi_d=-2G\theta h^2$$

Solving these equations, we have

(CONT.)

7.6 CONT.

$$\Phi_b=2.238G\theta h^2 \qquad \Phi_c=2.000G\theta h^2$$
$$\Phi_d=1.096G\theta h^2 \qquad \Phi_e=1.715G\theta h^2$$
$$\Phi_f=2.667G\theta h^2 \qquad \Phi_g=2.953G\theta h^2$$

The finite differences are the computed as shown in Table P7.6a. Thus,

$$\tau_A=\left(\frac{\partial\Phi}{\partial x}\right)_A=\frac{1}{h}\left[\Delta-\frac{\Delta^2}{2}+\frac{\Delta^3}{3}-\frac{\Delta^4}{4}\right]$$
$$=[2.238+\frac{1.523}{2}+\frac{0.093}{3}+\frac{0.0186}{4}]G\theta h$$

$$=3.035G\theta h=0.0129G\theta$$ ◄

Table P7.6a

y	$\Phi/G\theta h^2$	$\Delta/G\theta h^2$	$\Delta^2/G\theta h^2$	$\Delta^3/G\theta h^2$	$\Delta^4/G\theta h^2$
0	0	2.238	-1.523	0.093	-0.019
h	2.238	0.715	-1.430	-0.003	
2h	2.953	-0.715	-1.523		
3h	2.238	-2.238			
4h	0				

Alternatively, we construct the table as shown in Fig. P7.6b:

Table P7.6b

y	$\Phi/G\theta h^2$	$\Delta/G\theta h^2$	$\Delta^2/G\theta h^2$
0	2.953	-0.715	-1.523
h	2.238	-2.238	
2h	0		

Then we obtain,

$$\tau_A=\left(\frac{\partial\Phi}{\partial y}\right)_{y=2h}$$

$$=\frac{\Delta\Phi_0}{h}+\frac{3}{2}\frac{\Delta^2\Phi_0}{h}$$

$$=[-0.715-\frac{3}{2}(1.523)]G\theta h$$

$$=3.0G\theta h$$ ◄

This result is approximately the same as calculated before. Clearly now computations involved are reduced considerably.

7.7

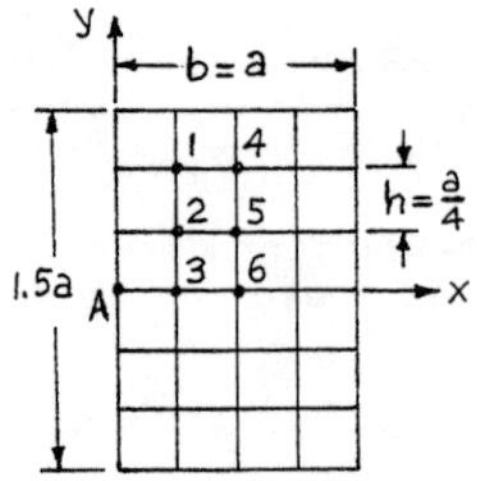

Applying Eq. (7.18) at nodes 1 to 6, we obtain

$$\begin{bmatrix} -4 & 1 & 0 & 1 & 0 & 0 \\ 1 & -4 & 1 & 0 & 1 & 0 \\ 0 & 2 & -4 & 0 & 0 & 1 \\ 2 & 0 & 0 & -4 & 1 & 0 \\ 0 & 2 & 0 & 1 & -4 & 1 \\ 0 & 0 & 2 & 0 & 2 & -4 \end{bmatrix} \begin{Bmatrix} \Phi_1 \\ \Phi_2 \\ \Phi_3 \\ \Phi_4 \\ \Phi_5 \\ \Phi_6 \end{Bmatrix} = -2G\theta h^2 \begin{Bmatrix} 1 \\ 1 \\ 1 \\ 1 \\ 1 \\ 1 \end{Bmatrix}$$

Solving,

$$\Phi_1 = 1.551G\theta h^2 \qquad \Phi_2 = 2.206G\theta h^2$$
$$\Phi_3 = 2.387G\theta h^2 \qquad \Phi_4 = 1.997G\theta h^2$$
$$\Phi_5 = 2.887G\theta h^2 \qquad \Phi_6 = 3.137G\theta h^2$$

Then, the differences are as shown in Table P7.7.

Table P7.7

x	$\Phi/G\theta h^2$	$\Delta/G\theta h^2$	$\Delta^2/G\theta h^2$	$\Delta^3/G\theta h^2$	$\Delta^4/G\theta h^2$
0	0	2.387	-1.637	0.137	-0.274
h	2.387	0.75	-1.50	-0.137	
2h	3.137	-0.75	-1.637		
3h	2.387	-2.387			
4h	0				

At point A:

$$\left(\frac{\partial\Phi}{\partial x}\right)_{x=0} = \frac{1}{h}\left(\Delta\Phi_0 - \frac{\Delta^2\Phi_0}{2} + \frac{\Delta^3\Phi_0}{3} - \frac{\Delta^4\Phi_0}{4}\right)$$

Substituting the first row of Table P7.7:

$$\left(\frac{\partial\Phi}{\partial x}\right)_A = \left[2.387 + \frac{1.637}{2} + \frac{0.137}{3} + \frac{0.274}{4}\right]G\theta h$$

or

$$\left(\frac{\partial\Phi}{\partial x}\right)_A = 0.83G\theta a = \tau_{max}$$ ◄

This differs 2.12% from the exact solution 0.848Gθa (Table 6.2).

7.8

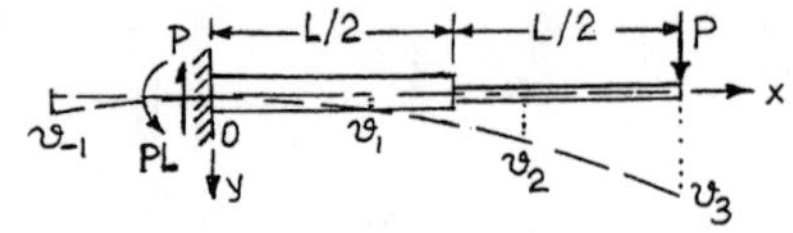

Boundary conditions yield,

$$v(0)=0; \quad v_0=0 \quad \text{and} \quad v'(0)=0; \quad v_1 = v_{-1}$$

Apply Eq. (7.27) at nodes 0, 1, and 2, respectively:

$$v_{-1} - 2v_0 + v_0 = \frac{h^2(PL)}{2EI}; \qquad v_1 = \frac{h^2}{4EI}PL$$
$$v_2 - 2v_1 + v_0 = \frac{h^2(2PL/3)}{2EI}; \qquad v_2 = \frac{5h^2}{6EI}PL$$
$$v_3 - 2v_2 + v_1 = \frac{h^2(PL/3)}{EI} \qquad \text{(a)}$$

Substituting v_1 and v_2 into Eq. (a) we obtain

$$v_3 = \frac{7}{32}\frac{PL^3}{EI}$$ ◄

7.9

Let $N = p_o L^4/EI$

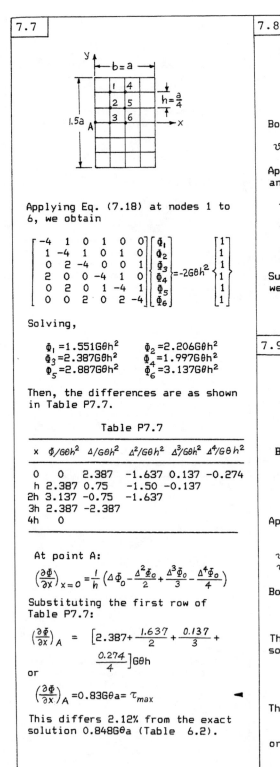

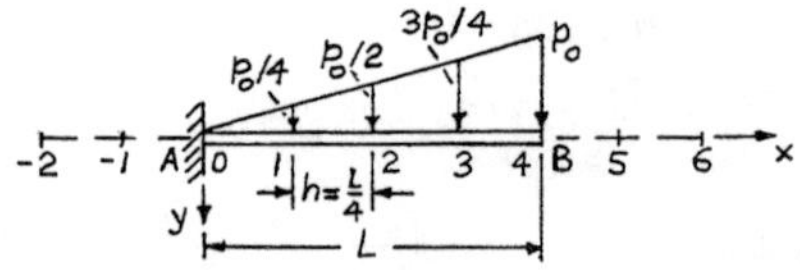

Boundary conditions at A:

$$v'(0)=0; \qquad v_1 = v_{-1}$$
$$v(0)=0; \qquad v_0 = 0$$
$$v^{IV}(0)=0; \qquad v_2 - 8v_1 + v_{-2} = 0$$

Apply Eq. (7.26) at 1 through 4:

$$7v_1 - 4v_2 + v_3 = (0.25)^5 N$$
$$-4v_1 + 6v_2 - 4v_3 + v_4 = (0.5)^9 N$$
$$v_1 - 4v_2 + 6v_3 - 4v_4 + v_5 = 3(0.5)^{10} N$$
$$v_2 - 4v_3 + 6v_4 - 4v_5 + v_6 = (0.25)^4 N$$

Boundary conditions at B:

$$v'''(L)=0; \qquad -v_2 + 2v_3 - 2v_5 + v_6 = 0$$
$$v''(L)=0; \qquad v_3 - 2v_4 + v_5 = 0$$

The foregoing six equations are solved to yield

$$v_1 = 0.010742N \qquad v_2 = 0.035156N$$
$$v_3 = 0.066406N \qquad v_4 = 0.099609N$$
$$v_5 = 0.132812N \qquad v_6 = 0.167969N$$ ◄

The errror is therefore

$$\frac{0.099609 - 0.091667}{0.091667} \times 100$$

or 8.7%.

7.10

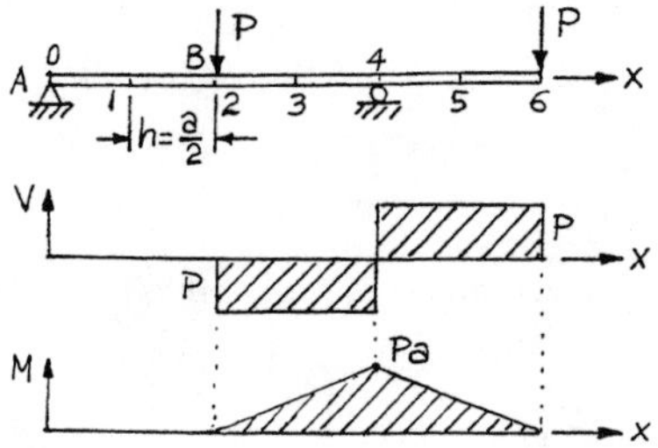

Equation (7.27) is applied at nodes 1, 2, 3, respectively:

$$0-2v_1+v_2=0; \qquad v_1=v_2/2$$
$$v_1-2v_2+v_3=0; \qquad v_2=2v_3/3$$
$$v_2-2v_3+0=(h^2/IE)(Pa/2)$$

or

$$v_3=-3Pah^2/8EI$$

Thus,

$$v_B=v_2=-\frac{1}{16}\frac{Pa^3}{EI}=-0.0625\frac{Pa^3}{EI}$$ ◄

and

$$v_1=-Pa^3/32EI$$

Slope at A is then

$$\theta_A=\frac{v_1-v_{-1}}{2h}=\frac{v_1}{h}=-\frac{1}{16}\frac{Pa^2}{EI}$$ ◄

We have v_B=0.08333Pa^3/EI as exact solution. Error is thus

$$\frac{0.08333-0.0625}{0.0833}\times 100 = 25\,\%$$

7.11

Boundary conditions yield

$$v(0)=0; \qquad v_0=0$$
$$v'(0)=0; \qquad v_1=v_{-1}$$

and

$$v''(L)=0=v_4-2v_3+v_2$$

or

$$v_4=2v_3-v_2 \qquad (1)$$
$$v'''(L)=0=v_5-2v_4+2v_2-v_1$$

or

$$v_5=4v_3-4v_2+v_1 \qquad (2)$$

Apply Eq. (7.26) at points 1, 2, and 3, with v_0=0 and $v_1=v_{-1}$:

$$v_3-4v_2+7v_1=\frac{pL^4}{162EI} \qquad (3)$$
$$v_4-4v_3-6v_2-4v_1=\frac{pL^4}{81EI} \qquad (4)$$
$$v_5-4v_4+6v_3-4v_2+v_1=\frac{pL^4}{81EI} \qquad (5)$$

Solving Eqs. (1) to (6),

$$v_3=7pL^4/54EI$$ ◄

and

$$v_1=4v_3/21 \qquad v_2=4v_3/7$$

7.12

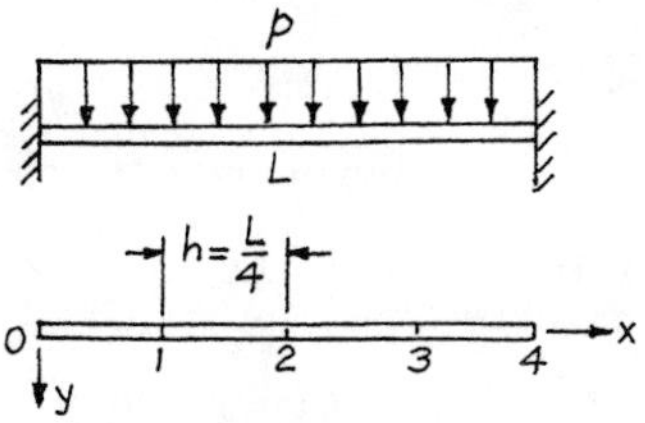

Boundary conditions are $v_0=v_4=0$ and from symmetry $v_1=v_3$.

Then, using Eq. (7.26) at nodes 1 and 2:

$$v_1+6v_1-4v_2+v_3=\frac{ph^4}{EI}$$
$$-4v_1+6v_2-4v_3=\frac{ph^4}{EI}$$

Solving,

$$v_2=\frac{ph^4}{EI}=\frac{p}{EI}\left(\frac{L}{4}\right)^4=0.00391\frac{pL^4}{EI}$$ ◄

The exact solution is

$$v_2=0.002604\frac{pL^4}{EI}$$

The slope is zero at nodes 0, 2, and 4. Thus,

$$\theta_1=\theta_3=\frac{v_2-v_0}{2h}=\frac{(0.00391)4}{2L}\frac{pL^4}{EI}$$
$$=0.078\frac{pL^3}{EI}$$ ◄

7.13

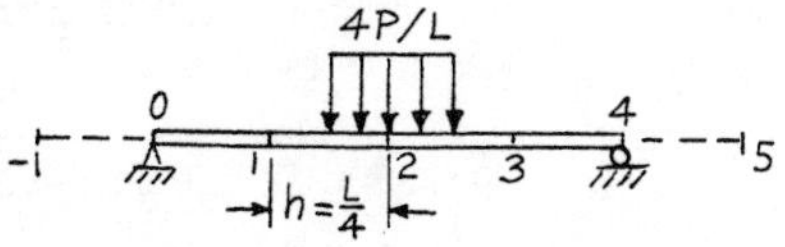

Equation (7.26) is applied at nodes 1, 2, 3 to obtain

$$v_3-4v_2+6v_1-4v_0+v_{-1}=0 \qquad (a)$$
$$v_4-4v_3+6v_2-4v_1+v_0$$
$$=\frac{L^4}{256}\frac{4P}{EIL}=\frac{PL^3}{64EI} \qquad (b)$$

$$v_5-4v_4+6v_3-4v_2+v_1=0 \qquad (c)$$

(CONT.)

7.13 CONT.

Boundary conditions are represented by

$$v_0 = v_4 = 0 \qquad v_1 = v_3 \qquad v_1 = -v_{-1} \qquad v_3 = -v_5 \qquad \text{(d)}$$

Note that due to the symmetry, Eqs. (a) and (c) are identical. Thus, by deleting one, using condititons (d), Eqs. (a) and (b) become

$$6v_1 = 4v_2 = 0 \qquad -4v_1 + 3v_3 = \frac{PL^3}{128EI}$$

Solving,

$$v_1 = 0.0156\,\frac{PL^3}{EI} \qquad v_{max} = v_2 = \frac{PL^3}{42.7\,EI} \quad \blacktriangleleft$$

The maximum slope, from Eq. (a) of Sec. 7.7:

$$\theta_{max} = \theta_0 = \frac{v_1 - v_{-1}}{2h} = \frac{v_1}{h} = 0.0624\,\frac{PL^2}{EI} \quad \blacktriangleleft$$

7.14

Note that analyses of the frame shown in Fig. P7.14 and the continuous beam (following sketch) proceed in an identical fashion.

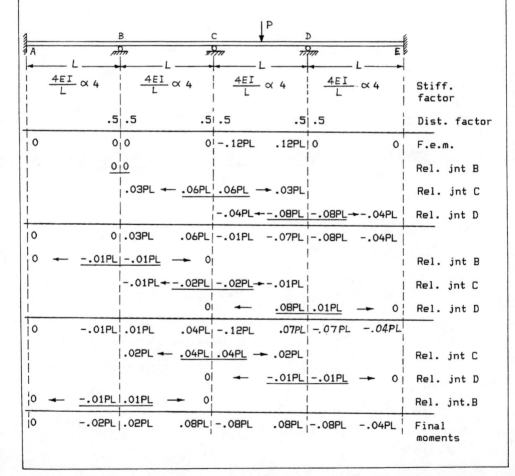

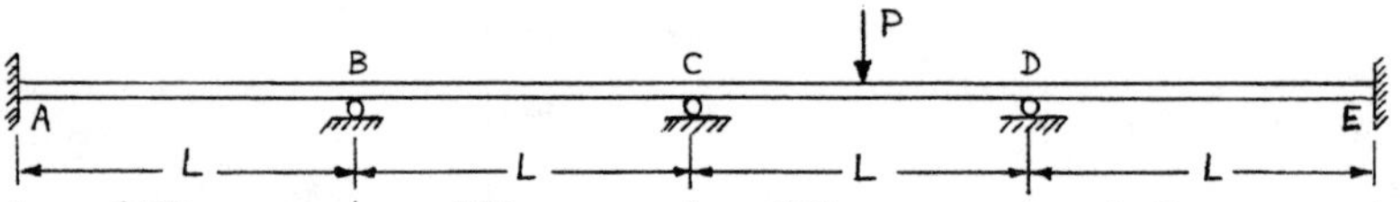

A	B	B	C	C	D	D	E	
$\frac{4EI}{L} \propto 4$		$\frac{4EI}{L} \propto 4$		$\frac{4EI}{L} \propto 4$		$\frac{4EI}{L} \propto 4$		Stiff. factor
	.5	.5	.5	.5	.5	.5		Dist. factor
0	0	0	0	−.12PL	.12PL	0	0	F.e.m.
	0	0						Rel. jnt B
		.03PL ←	.06PL	.06PL →	.03PL			Rel. jnt C
				−.04PL ←	−.08PL	−.08PL →	−.04PL	Rel. jnt D
0	0	.03PL	.06PL	−.01PL	−.07PL	−.08PL	−.04PL	
0 ←	−.01PL	−.01PL →	0					Rel. jnt B
		−.01PL ←	−.02PL	−.02PL →	−.01PL			Rel. jnt C
			0 ←		.08PL	.01PL →	0	Rel. jnt D
0	−.01PL	.01PL	.04PL	−.12PL	.07PL	−.07PL	−.04PL	
		.02PL ←	.04PL	.04PL →	.02PL			Rel. jnt C
			0 ←		−.01PL	−.01PL →	0	Rel. jnt D
0 ←	−.01PL	.01PL →	0					Rel. jnt.B
0	−.02PL	.02PL	.08PL	−.08PL	.08PL	−.08PL	−.04PL	Final moments

7.15

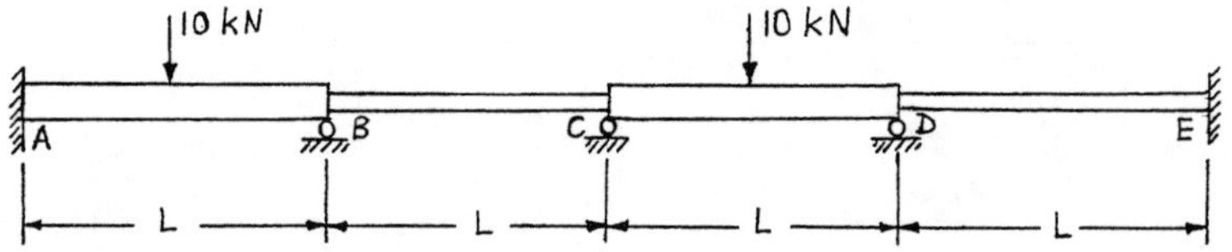

$\frac{4EI_0}{L} \propto 4$		$\frac{4EI_0}{L} \propto 2$		$\frac{4EI_0}{L} \propto 4$		$\frac{4EI_0}{L} \propto 2$		Stiff. factor
0.667	0.667	0.333	0.333	0.667	0.667	0.333	0.333	Dist. factor
-1.25	1.25	0	0	-1.25	1.25	0	0	F.e.m.
0.83 →	0.42					0.21 ←	0.42	Rel. jnt A
-0.56 ←	-1.11	-0.56 →	-0.28					Rel. jnt B
		0.26 ←	0.51	1.02 →	0.51			Rel. jnt C
				-0.66 ←	-1.31	-0.66 →	-0.33	Rel. jnt D
-0.98	0.56	-0.30	0.23	-0.89	0.45	-0.45	0.09	
0.59 →	0.30					0.15 ←	0.30	Rel jnt. A
-0.19 ←	-0.37	-0.19 →	-0.09					Rel. jnt B
		0.13 ←	0.25	0.50 →	0.25			Rel. jnt C
				-0.13 ←	-0.27	-0.13 →	-0.07	Rel. jnt D
-0.58	0.49	-0.36	0.39	-0.52	0.43	-0.43	0.32	
0.17 →	0.09					0.04 —	0.09	Rel. jnt A
-0.07 ←	-0.15	-0.07 →	-0.04					Rel. jnt B
		0.03 ←	0.06	0.11 →	0.06			Rel. jnt C
				-0.03 ←	-0.07	-0.03 →	-0.02	Rel. jnt D
-0.48	0.43	-0.40	0.41	-0.44	0.42	-0.42	0.39	
0.06 →	0.03					0.02 ←	0.03	Rel. jnt A
-0.02 ←	-0.04	-0.02 →	-0.01					Rel. jnt B
		0.01 ←	0.01	0.03 →	0.01			Rel. jnt C
				-0.01 ←	-0.02	-0.01 →	-0.01	Rel. jnt D
-0.44	0.42	-0.41	0.41	-0.42	0.41	-0.41	0.41	
0.02 →	0.01					0.00 ←	0.01	Rel. jnt A
0.00 ←	-0.01	-0.01 →	0.00					Rel. jnt B
		0.00 ←	0.00	0.01 →	0.00			Rel. jnt C
				0.00 ←	0.00	0.00 →	0.00	Rel. jnt D
-0.42	0.42	-0.42	0.41	-0.41	0.41	-0.41	0.42	
-0.42L	0.42L	-0.42L	0.41L	-0.41L	0.41L	-0.41L	0.42L	Final moments

Note: An "L" is contained in all the moments; it is omitted until the final answer, as shown in the preceding solution.

7.16

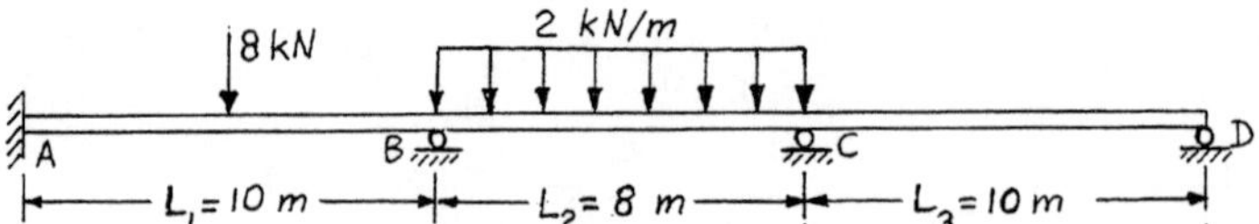

A		B	B		C	C	D	
$\frac{4EI_1}{L_1} = \frac{2}{5}EI_1$			$\frac{4EI_2}{L_2} = \frac{1}{2}EI_1$			$\frac{3EI_1}{L_3} = \frac{3}{20}EI_1$		Stiff. factor
		0.444	0555		0.769	0.231		Dist. factor
-10.0		10.0	-10.6		10.6	0		F.e.m.
0.1	←	0.3	0.3	→	0.1			Rel. jnt B
			-4.1	←	-8.2	-2.5		Rel. jnt C
0.9	←	1.8	2.3	→	1.0			Rel. jnt B
			-0.4	←	-0.8	-0.2		Rel. jnt C
0.1	←	0.2	0.2	→	0.1			Rel. jnt B
			0	←	-0.1	0		Rel. jnt C
-8.9		12.3	-12.3		2.7	-2.7	0	Final moments

7.17

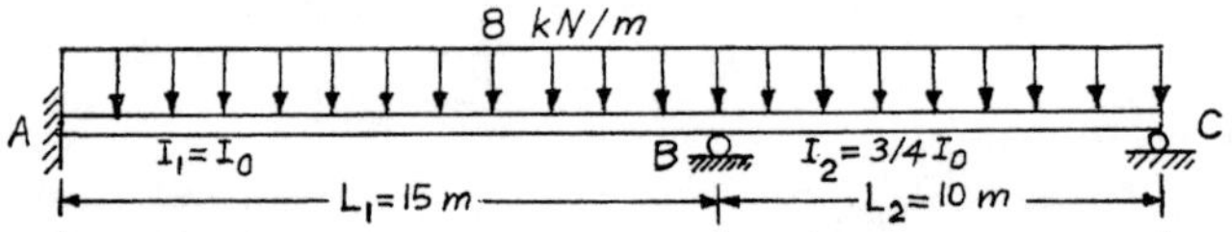

A		B	B		C	
$\frac{4EI_1}{L_1} \propto \frac{4}{15}$			$\frac{3EI_2}{L_2} \propto \frac{9}{40}$			Stiff. factor
		0.542	0.458		1.0	Dist. factor
-150.00		150.00	-66.67		66.67	F.e.m.
-22.58	←	-45.16	-38.17	→	0	Rel. jnt B
			-33.33	←	-66.67	Rel. jnt C
-172.58		104.84	138.17		0	
9.03	←	18.06	15.27			Rel. jnt B
-163.55		122.90	-122.90		0	Final moments

7.18

(a) Refer to Fig. 7.18a:

$$Q_j = \tfrac{1}{2}p_j\, h_1\, t + \tfrac{1}{3}\,\frac{p_m - p_j}{2}\, h_1\, t = h_1\, t(2p_j + p_m)/6$$

$$Q_m = \tfrac{1}{2}p_j\, h_1\, t + \tfrac{2}{3}\,\frac{p_m - p_j}{2}\, h_1\, t = h_1\, t(2p_m + p_j)/6$$

(b) Refer to Fig. 7.18b:

$$\tau_{xy} = \frac{PQ}{It} = \frac{3P}{4th^3}(h^2 - y^2) \qquad \text{(a)}$$

Substituting Eq. (a) into the given expression for Q_m:

$$Q_m = \frac{1}{y_m - y_j}\int_{y_j}^{y_m} \frac{3P}{4h^3}(h^2 - y^2)(y - y_j)\,dy$$

(CONT.)

(7.18 CONT.)

$$= \frac{1}{y_m - y_j}\,\frac{3P}{4h}\left[-y_j\left(y - \frac{y^3}{3h^2}\right)\Big|_{y_j}^{y_m} + \left(\frac{y^2}{2} - \frac{y^4}{4h^2}\right)\Big|_{y_j}^{y_m}\right]$$

This leads to the first of Eqs. (7.57). Similarly,

$$Q_j = \int_{y_j}^{y_m} \tau_{xy}\, t\, dy - Q_m \qquad \text{(b)}$$

Substituting Eq. (a) and Q_m (from the first of Eqs. 7.57), Eq. (b) leads to the second of Eqs. (7.57).

7.19

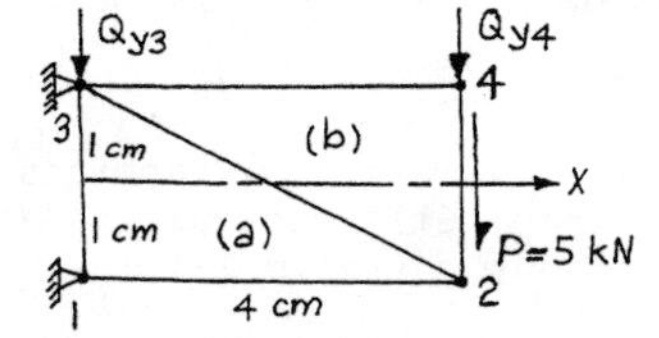

Body force effects:

$$\{Q\}_a^b = \frac{1}{3}At\{Q_{x1}, Q_{x2}, Q_{x3}, Q_{y1}, Q_{y2}, Q_{y3}\}$$

$$= \frac{1}{3}(4\times0.3)\{0,\ 0,\ 0,\ -0.077,\ -0.077,\ -0.077\}\ \text{N}$$

or

$$\{Q\}_a^b = \{0,\ 0,\ 0,\ 0,\ -.0308,\ -.0308,\ -.0308,\ -.0308,\ 0\}\ \text{N}$$

Similarly,

$$\{Q\}_b^b = \{0,\ 0,\ 0,\ 0,\ 0,\ -.0308,\ -.0308,\ -0308\}\ \text{N}$$

Hence,

$$\{Q\}^b = \{Q\}_a^b + \{Q\}_b^b$$

or

$$\{Q\}^b = \{0,\ 0,\ 0,\ 0,\ -.0308,\ -.0616,\ -.0616,\ -.0308\}\ \text{N}$$

Effect of shear force, P:

From Example 7.9,

$$\{Q\}^P = \{0,\ 0,\ 0,\ 0,\ -2500,\ -2500\}\ \text{N}$$

Surface traction effects, p:
Total load (4x0.3)700=840 N is equally divided between nodes 3 and 4. Thus,

$$\{Q\}^p = \{0,0,0,0,0,0,-420,-420\}\ \text{N}$$

Thermal strain effects:

We have $\varepsilon_0 = 12(10^{-6})50 = 600\mu$ and

$$[B]_a = \frac{1}{2A}\begin{bmatrix} b_1 & b_2 & b_3 & 0 & 0 & 0 \\ 0 & 0 & 0 & a_1 & a_2 & a_3 \\ a_1 & a_2 & a_3 & b_1 & b_2 & b_3 \end{bmatrix}$$

$$= \frac{1}{8}\begin{bmatrix} -2 & 2 & 0 & 0 & 0 & 0 \\ 0 & 0 & 0 & -4 & 0 & 4 \\ -4 & 0 & 4 & -2 & 2 & 0 \end{bmatrix}$$

Hence,

$$\{Q\}_a^t = [B]_a^T[D]\{\varepsilon_0\}(At)$$

$$= \frac{1}{8}\begin{bmatrix} -2 & 0 & -4 \\ 2 & 0 & 0 \\ 0 & 0 & 4 \\ 0 & -4 & -2 \\ 0 & 0 & 2 \\ 0 & 4 & 0 \end{bmatrix} \frac{2(10^7)}{0.91}\begin{bmatrix} 1 & .3 & 0 \\ .3 & 1 & 0 \\ 0 & 0 & .35 \end{bmatrix}\begin{Bmatrix} 600\mu \\ 600\mu \\ 0 \end{Bmatrix}(1.2)$$

(CONT.)

7.19 CONT.

After multiplication, this gives

$$\{Q\}_a^t = \{-5142.85,\ 5142.85,\ 0,\ -10285.7,\ 0,\ 10285.7\}\ \text{N}$$

or

$$\{Q\}_a^t = \{-5142.85,\ 5142.85,\ 0,\ 0,\ -10285.7,\ 0,\ 10285.7,\ 0\}\ \text{N}$$

Similarly,

$$[B]_b = \frac{1}{2A}\begin{bmatrix} b_2 & b_4 & b_3 & 0 & 0 & 0 \\ 0 & 0 & 0 & a_2 & a_4 & a_3 \\ a_2 & a_4 & a_3 & b_2 & b_4 & b_3 \end{bmatrix}$$

$$= \frac{1}{8}\begin{bmatrix} 0 & 2 & -2 & 0 & 0 & 0 \\ 0 & 0 & 0 & -4 & 4 & 0 \\ -4 & 4 & 0 & 0 & 2 & -2 \end{bmatrix}$$

Thus,

$$\{Q\}^t = [B]_b^T[D]\{\varepsilon_0\}(At)$$

$$= \{0,\ 5142.85,\ -5142.85,\ -10285.7,\ 10285.7,\ 0\}\ \text{N}$$

or

$$\{Q\}_b^t = \{0,\ 0,\ -5142.85,\ 5142.85,\ 0,\ -10285.7,\ 0,\ 10285.7\}\ \text{N}$$

Hence,

$$\{Q\}^t = \{Q\}_a^t + \{Q\}_b^t$$

or

$$\{Q\}^t = \{-5142.85,\ 5142.85,\ -5142.85,\ 5142.85,\ -10285.7,\ -10285.7,\ 10285.7,\ 10285.7\}\ \text{N}$$

The system nodal force matrix:

$$\{Q\} = \{Q\}^b + \{Q\}^P + \{Q\}^p + \{Q\}^t$$

$$= \{-5142.85,\ 5142.85,\ -5142.85,\ 5142.85,\ -10285.7,\ -12785.76,\ 986.64,\ 7365.67\}\ \text{N}$$

System equation:

$$\{Q\} = [K]\{0,\ u_2,\ 0,\ u_4,\ 0, v_2,\ 0, v_4\}$$

where

[K] is given by Eq. (n) of Example 7.9.

Since we have only 4 unknown quantities u_2, u_4, v_2, v_4 (and 8 equations are available), there are redundant equations. Examination of these system of equations shows that:

[K] is reduced by crossing out; row 1 and column 1 for $u_1=0$, row 3 and column 3 for $u_3=0$, row 5 and column 5 for $v_1=0$, row 7 and column 7 for $v_3=0$.

{Q} is reduced by crossing out

$$Q_{x1}, Q_{x3}, Q_{y1}, Q_{y3}$$

for

$$u_1 = u_3 = v_1 = v_3 = 0$$

(CONT.)

7.19 CONT.

Thus, from the reduced equations, we obtain

$$\begin{Bmatrix} u_2 \\ u_4 \\ v_2 \\ v_4 \end{Bmatrix} = 10^{-6} \begin{bmatrix} .429 & .180 & .252 & .247 \\ .180 & .483 & -.256 & -.351 \\ .252 & -.256 & 1.366 & 1.373 \\ .247 & -.351 & 1.373 & 1.546 \end{bmatrix}$$

$$\times\{5142.85, 5142.85, -12785.76, 7365.67\}$$

or

$$\begin{Bmatrix} u_2 \\ u_4 \\ v_2 \\ v_4 \end{Bmatrix} = \begin{Bmatrix} 1729 \\ 4098 \\ -7302 \\ -6702 \end{Bmatrix} 10^{-6} \text{ cm}$$

7.20

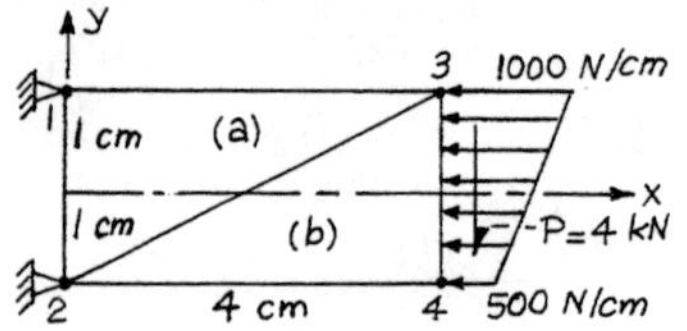

We have

$$n = 2.1(10^7)/7(10^6) = 3$$

$$m = 2.8(10^7)/7(10^6) = 0.4$$

$$[D] = \frac{7(10^6)}{1-3(0.1)^2}\begin{bmatrix} 3 & 0.3 & 0 \\ 0.3 & 1 & 0 \\ 0 & 0 & 0.39 \end{bmatrix}$$

Then

$$[D^*] = t[D]/4A = (2/7)[D]/16$$

$$= 0.13(10^6)\begin{bmatrix} 3 & 0.3 & 0 \\ 0.3 & 1 & 0 \\ 0 & 0 & 0.39 \end{bmatrix} \qquad \text{(a)}$$

Element a:
Equations (7.46), with reference to the preceding figure yield

$$a_i = a_2 = -4 \qquad b_i = b_2 = 0$$
$$a_j = a_3 = 0 \qquad b_j = b_3 = 2 \qquad \text{(b)}$$
$$a_m = a_1 = 4 \qquad b_m = b_1 = -2$$

Substituting Eqs. (a) and (b) into Eqs. (7.54), we obtain (in 10^6):

$k_{uu,11} = 0.13[3(4)+0.39(16)+0] = 2.37$
$k_{uu,12} = 0.13[0+0.39(-16)+0] = -0.81$
$k_{uu,13} = 0.13[3(-4)+0+0] = -1.56$
$k_{uu,21} = 0.13[0+0.39(-16)+0] = -0.81$
$k_{uu,22} = 0.13[0+0.39(16)+0] = 0.81$
$k_{uu,23} = 0.13[0+0+0] = 0$
$k_{uu,31} = 0.13[3(-4)+0+0] = -1.56$
$k_{uu,32} = 0.13[0+0+0] = 0$
$k_{uu,33} = 0.13[3(4)+0+0] = 1.56$

(CONT.)

7.20 CONT.

Similarly, remaining matrices are determined. Stiffness matrix for the element a (in 10^6) is:

$$[k]_a = \begin{bmatrix} 2.37 & -.81 & -1.56 & -.72 & .31 & .41 \\ -.81 & .81 & 0 & .41 & 0 & -.41 \\ -1.56 & 0 & 1.56 & .31 & -.31 & 0 \\ -.72 & .41 & .31 & 2.28 & -2.08 & -.20 \\ .31 & 0 & -.31 & -2.08 & 2.08 & 0 \\ .41 & -.41 & 0 & -.20 & 0 & .20 \end{bmatrix}$$

or

$$[k]_a = \begin{bmatrix} 2.37 & -.81 & -1.56 & 0 & .72 & .31 & .41 & 0 \\ -.81 & .81 & 0 & 0 & .41 & 0 & -.41 & 0 \\ -1.56 & 0 & 1.56 & 0 & .31 & -.31 & 0 & 0 \\ 0 & 0 & 0 & 0 & 0 & 0 & 0 & 0 \\ -.72 & .41 & .31 & 0 & 2.28 & -2.08 & -.20 & 0 \\ .31 & 0 & -.31 & 0 & -2.08 & 2.08 & 0 & 0 \\ .41 & -.41 & 0 & 0 & -.20 & 0 & .20 & 0 \\ 0 & 0 & 0 & 0 & 0 & 0 & 0 & 0 \end{bmatrix}$$

Element b:
Equations (7.46), referring to the preceding figure, now give

$$a_i = a_2 = 0 \qquad b_i = b_2 = -2$$
$$a_j = a_4 = -4 \qquad b_j = b_4 = 2 \qquad \text{(c)}$$
$$a_m = a_3 = 4 \qquad b_m = b_3 = 0$$

Introducing Eqs. (a) and (c) into Eqs. (7.54), we obtain (in 10):

$k_{uu,22} = 0.13[3(4)+0+0] = 1.56$
$k_{uu,23} = 0.13[0+0+0] = 0$
$k_{uu,24} = 0.13[3(-4)+0+0] = -1.56$

The remaining terms are obtained in a like manner. The stiffness matrix for the element b (in 10^6):

$$[k]_b = \begin{bmatrix} 1.56 & 0 & -1.56 & 0 & -.31 & .31 \\ 0 & .81 & -.81 & -.41 & 0 & .41 \\ -1.56 & -.81 & 2.37 & .41 & .31 & -.72 \\ 0 & -.41 & .41 & .2 & 0 & -.2 \\ -.31 & 0 & .31 & 0 & 2.08 & -2.08 \\ .31 & .41 & -.72 & -.2 & -2.08 & 2.28 \end{bmatrix}$$

or

$$[k]_b = \begin{bmatrix} 0 & 0 & 0 & 0 & 0 & 0 & 0 & 0 \\ 0 & 1.56 & 0 & -1.56 & 0 & 0 & -.31 & .31 \\ 0 & 0 & .81 & -.81 & 0 & -.41 & 0 & .41 \\ 0 & -1.56 & -.81 & 2.37 & 0 & .41 & .31 & -.72 \\ 0 & 0 & 0 & 0 & 0 & 0 & 0 & 0 \\ 0 & 0 & -.41 & .41 & 0 & .20 & 0 & -.20 \\ 0 & -.31 & 0 & .31 & 0 & 0 & 2.08 & -2.08 \\ 0 & .31 & .41 & -.72 & 0 & -.20 & -2.08 & 2.28 \end{bmatrix}$$

The system matrix is found by addition of the matrices of the elements a and b:

$$[K] = [k]_a + [k]_b$$

That is (in 10^6),

(CONT.)

7.20 CONT.

$$[K]=\left[\begin{array}{cccc|cccc} 2.37 & -.81 & -1.56 & 0 & -.72 & .31 & .41 & 0 \\ -.81 & 2.37 & 0 & -1.56 & .41 & 0 & .72 & .31 \\ -1.56 & 0 & 2.37 & -.81 & .31 & -.72 & 0 & .41 \\ 0 & -1.56 & -.81 & 2.37 & 0 & .41 & .31 & -.72 \\ \hline -.72 & .41 & .31 & 0 & 2.28 & -2.08 & -.20 & 0 \\ .31 & 0 & -.72 & .41 & -2.08 & 2.28 & 0 & -.20 \\ .41 & -.72 & 0 & .31 & -.20 & 0 & 2.28 & -2.08 \\ 0 & .31 & .41 & -.72 & 0 & -0.20 & -2.08 & 2.28 \end{array}\right]$$

Nodal force matrix:

Equations (7.56) lead to

$Q_j=Q_{x4}=-2[2(500)+1000]/6=-667$ N

$Q_m=Q_{x3}=-2[2(1000)+500]/6=-833$ N

Due to the shear, we also have

$Q_{y3}=-2000$ N, $Q_{y4}=-2000$ N

Thus,

$\{Q\}=\{0,\ 0,\ -833,\ -667,\ 0,\ 0,\ -2000,\ -2000\}$ N

Nodal displacement matrix:

$\{\delta\}=\{0,\ 0,\ u_3, u_4,\ 0,\ 0,\ v_3, v_4\}$

The system equation:

The redundant equations are eliminated by crossing out the 1st, 2nd, 5th, and 6th rows and columns. This leaves

$$\begin{Bmatrix} -833 \\ -667 \\ -2000 \\ -2000 \end{Bmatrix}=10^6\begin{bmatrix} 2.37 & -.81 & 0 & .41 \\ -.81 & 2.37 & .31 & -.72 \\ 0 & .31 & 2.28 & -2.08 \\ .41 & -.72 & -2.08 & 2.28 \end{bmatrix}\begin{Bmatrix} u_3 \\ u_4 \\ v_3 \\ v_4 \end{Bmatrix}$$

Solving,

$u_3=0.001226$ cm, $u_4=-0.002324$ cm

$v_3=-0.013092$ cm, $v_4=-0.013742$ cm

Stress in element a:

The strain matrix is

$$\begin{Bmatrix} \varepsilon_x \\ \varepsilon_y \\ \gamma_{xy} \end{Bmatrix}=\frac{1}{2A}\begin{bmatrix} b_i & b_j & b_m & 0 & 0 & 0 \\ 0 & 0 & 0 & a_i & a_j & a_m \\ a_i & a_j & a_m & b_i & b_j & b_m \end{bmatrix}\begin{Bmatrix} u_i \\ u_j \\ u_m \\ v_i \\ v_j \\ v_m \end{Bmatrix}$$

$$=\frac{10^{-6}}{8}\begin{bmatrix} 0 & 2 & -2 & 0 & 0 & 0 \\ 0 & 0 & 0 & -4 & 0 & 4 \\ -4 & 0 & 4 & 0 & 2 & -2 \end{bmatrix}\begin{Bmatrix} 0 \\ 1226 \\ 0 \\ 0 \\ -13{,}092 \\ 0 \end{Bmatrix}$$

$$=\begin{Bmatrix} 307 \\ 0 \\ -3273 \end{Bmatrix}\mu$$

(CONT.)

7.20 CONT.

Then,

$$\begin{Bmatrix} \sigma_x \\ \sigma_y \\ \tau_{xy} \end{Bmatrix}_a=[D]\{\varepsilon\}_a$$

$$=\frac{70(10^9)}{0.97}\begin{bmatrix} 3 & .3 & 0 \\ .3 & 1 & 0 \\ 0 & 0 & .39 \end{bmatrix}\begin{Bmatrix} 307 \\ 0 \\ -3273 \end{Bmatrix}10^{-6}$$

$=\{66.46,\ 6.65,\ -92.12\}$ MPa

Stress in element b:

The strain matrix is

$$\begin{Bmatrix} \varepsilon_x \\ \varepsilon_y \\ \gamma_{xy} \end{Bmatrix}_b=\frac{10^{-6}}{8}\begin{bmatrix} -2 & 2 & 0 & 0 & 0 & 0 \\ 0 & 0 & 0 & 0 & -4 & 4 \\ 0 & -4 & 4 & -2 & 2 & 0 \end{bmatrix}\begin{Bmatrix} 0 \\ 2324 \\ -1226 \\ 0 \\ -13{,}742 \\ 13{,}092 \end{Bmatrix}$$

$=\{-581,\ 325,\ -1661\}\mu$

Thus,

$$\begin{Bmatrix} \sigma_x \\ \sigma_y \\ \tau_{xy} \end{Bmatrix}_b=\frac{70(10^3)}{0.97}\begin{bmatrix} 3 & .3 & 0 \\ .3 & 1 & 0 \\ 0 & 0 & .39 \end{bmatrix}\begin{matrix} -581 \\ 325 \\ -1661 \end{matrix}\begin{Bmatrix} -581 \\ 325 \\ -1661 \end{Bmatrix}$$

$=\{-118.75,\ 10.88,\ -46.75\}$ MPa

7,21

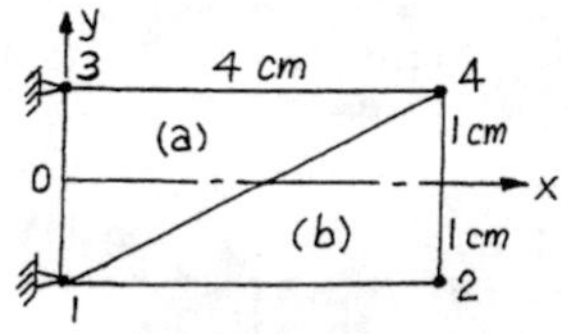

Stiffness matrix of element a:

Let $i=1$, $j=4$, $m=3$. Then,

$x_1=0$ $\quad x_4=4$ $\quad x_3=0$

$y_1=-1$ $\quad y_4=4$ $\quad y_3=1$

Equations (7.46) give

$a_1=0-4=-4$ $\quad b_1=1-1=0$

$a_4=0-0=0$ $\quad b_4=1+1=2$

$a_3=4-0=4$ $\quad b_3=-1-1=-2$

We have

$$[D^*]=\frac{10^6}{8}\begin{bmatrix} 3.33 & 0.99 & 0 \\ 0.99 & 3.3 & 0 \\ 0 & 0 & 1.16 \end{bmatrix}$$

Equations (7.54) in 10^6 are thus,

$k_{uu,11}=[0+1.16(16)]/8=2.32$

$k_{uu,14}=[0+0+0]/8=0$

$k_{uu,13}=[0+1.16(-16)+0]/8=-2.32$

$k_{uu,44}=[3.3(4)+0+0]/8=1.65$

$k_{uu,43}=[3.3(-4)+0+0]/8=-1.65$

$k_{uu,33}=[3.3(4)+0+0]/8=3.97$

These can be written as follows:

(CONT.)

7.21 CONT.

$$k_{uu} = \begin{bmatrix} 2.32 & -2.32 & 0 \\ -2.32 & 3.97 & -1.65 \\ 0 & -1.65 & 1.65 \end{bmatrix} (10^6)$$

Similarly, we find submatrices k_{vv} and k_{uv}. In so doing and after assembling these matrices, we obtain the stiffness matrix for the element a (in 10^6):

$$[k]_a = \left[\begin{array}{ccc|ccc} 2.32 & -2.32 & 0 & 0 & 1.16 & -1.16 \\ -2.32 & 3.97 & -1.65 & 0.99 & -2.15 & 1.16 \\ 0 & -1.65 & 1.65 & -0.99 & 0.99 & 0 \\ \hline 0 & 0.99 & -0.99 & 6.6 & -6.6 & 0 \\ 1.16 & -2.15 & 0.99 & -6.6 & 7.18 & -0.58 \\ -1.16 & 1.16 & 0 & 0 & -0.58 & 0.58 \end{array}\right]$$

Stiffness matrix for element b:

Let i=1, j=2, m=4. Then

$a_1 = 4-4 = 0$ $\quad b_1 = -1-1 = -2$

$a_2 = 0-4 = -4$ $\quad b_2 = 1+1 = 2$

$a_4 = 4-0 = 4$ $\quad b_4 = -1+1 = 0$

Then, Eqs. (7.54) yield (in 10^6):

$k_{uu,11} = [3.3(4)+0+0]/8 = 1.65$

$k_{uu,12} = [3.3(-4)+0+0]/8 = -1.65$

$k_{uu,14} = [0+0+0]/8 = 0$

$k_{uu,22} = [3.3(4)+1.16(16)+0]/8 = 3.97$

$k_{uu,24} = [0+1.16(-16)+0]/8 = -2.32$

$k_{uu,44} = [0+1.16(6)+0]/8 = 2.32$

or

$$k_{uu} = \begin{bmatrix} 1.65 & -1.65 & 0 \\ -1.65 & 3.97 & -2.32 \\ 0 & -2.32 & 2.32 \end{bmatrix} (10^6)$$

Similarly, we obtain submatrices k_{vv} and k_{uv}. In so doing and after assembling these matrices, we determine the stiffness matrix for the element b (in 10^6):

$$[k]_b = \left[\begin{array}{ccc|ccc} 1.65 & -1.65 & 0 & 0 & 0.99 & -0.99 \\ -1.65 & 3.97 & -2.32 & 1.16 & -2.15 & 0.99 \\ 0 & -2.32 & 2.32 & -1.16 & 1.16 & 0 \\ \hline 0 & 1.16 & -1.16 & 0.58 & -0.58 & 0 \\ 0.99 & -2.15 & 1.16 & -0.58 & 7.18 & -6.6 \\ -0.99 & 0.99 & 0 & 0 & -6.6 & 6.6 \end{array}\right]$$

Prior to addition: the 2nd and 6th rows and columns of zeros are added to the matrix $[k]_a$; the 3rd and 7th rows and columns of zeros are added to the matrix $[k]_b$.

The system matrix:

$$[K] = [k]_a + [k]_b$$

is then (in 10^6):

(CONT.)

7.21 CONT.

$$[K] = \left[\begin{array}{cccc|cccc} 3.97 & -1.65 & -2.32 & 0 & 0 & 0.99 & 1.16 & -2.15 \\ -1.65 & 3.97 & 0 & -2.32 & 1.16 & -2.15 & 0 & 0.99 \\ -2.32 & 0 & 3.97 & -1.65 & 0.99 & 0 & -2.15 & 1.16 \\ 0 & -2.32 & -1.65 & 3.97 & -2.15 & 1.16 & 0.99 & 0 \\ \hline 0 & 1.16 & 0.99 & -2.15 & 7.18 & -0.58 & -6.6 & 0 \\ 0.99 & -2.15 & 0 & 1.16 & -0.58 & 7.18 & 0 & -6.6 \\ 1.66 & 0 & -2.15 & 0.99 & -6.6 & 0 & 7.18 & -0.58 \\ -2.15 & 0.99 & 1.16 & 0 & 0 & -6.6 & -0.58 & 7.18 \end{array}\right]$$

The force-displacement relation, Eq. (q) of Example 7.9 becomes

$$\begin{Bmatrix} 0 \\ 0 \\ -1.2 \\ -2.5 \end{Bmatrix} = \begin{bmatrix} 3.97 & -2.32 & -2.15 & 0.99 \\ -2.32 & 3.97 & 1.16 & 0 \\ -2.15 & 1.16 & 7.18 & -6.6 \\ 0.99 & 0 & -6.6 & 7.18 \end{bmatrix} \begin{Bmatrix} u_2 \\ u_4 \\ v_2 \\ v_4 \end{Bmatrix} 10^3$$

This yields

$$\begin{Bmatrix} u_2 \\ u_4 \\ v_2 \\ v_4 \end{Bmatrix} = 10^{-6} \begin{bmatrix} 483.2 & 179.9 & 357.9 & 225.8 \\ 178.6 & 429 & -246.8 & 251.7 \\ 357.9 & -247 & 1546 & -1373 \\ 255.8 & -251.7 & -1373 & -1366 \end{bmatrix} \begin{Bmatrix} 0 \\ 0 \\ -2.5 \\ -2.5 \end{Bmatrix}$$

$$= \begin{Bmatrix} -1534.35 \\ 1246.325 \\ -433 \\ 17.125 \end{Bmatrix} (10^{-6}) \text{ cm}$$

Stresses in element b:

$$\begin{Bmatrix} \varepsilon_x \\ \varepsilon_y \\ \gamma_{xy} \end{Bmatrix}_b = \frac{1}{2A} \begin{bmatrix} b_1 & b_2 & b_4 & 0 & 0 & 0 \\ 0 & 0 & 0 & a_1 & a_2 & a_4 \\ a_1 & a_2 & a_4 & b_1 & b_2 & b_4 \end{bmatrix} \begin{Bmatrix} u_1 \\ u_2 \\ u_4 \\ v_1 \\ v_2 \\ v_4 \end{Bmatrix}$$

$$= \frac{10^{-6}}{8} \begin{bmatrix} -2 & 2 & 0 & 0 & 0 & 0 \\ 0 & 0 & 0 & 0 & -4 & 4 \\ 0 & -4 & 4 & -2 & 2 & 0 \end{bmatrix} \begin{Bmatrix} 0 \\ -1534.35 \\ 1246.325 \\ 0 \\ -433 \\ 17.125 \end{Bmatrix}$$

$= \{-383.6,\ 225,\ 1282\}\ \mu$

Then,

$$\begin{Bmatrix} \sigma_x \\ \sigma_y \\ \tau_{xy} \end{Bmatrix}_b = \frac{200(10^3)}{0.91} \begin{bmatrix} 1 & 0.3 & 0 \\ 0.3 & 1 & 0 \\ 0 & 0 & 0.35 \end{bmatrix} \begin{Bmatrix} -383.6 \\ 225 \\ 1282 \end{Bmatrix}$$

$= \{-69.47,\ 24.16,\ 98.6\}$ MPa

Stresses in element a:

$$\begin{Bmatrix} \varepsilon_x \\ \varepsilon_y \\ \gamma_{xy} \end{Bmatrix}_a = \frac{10^{-6}}{8} \begin{bmatrix} 0 & -2 & 2 & 0 & 0 & 0 \\ 0 & 0 & 0 & -4 & 4 & 0 \\ -4 & 4 & 0 & 0 & -2 & 2 \end{bmatrix} \begin{Bmatrix} 0 \\ 0 \\ 1246.325 \\ 0 \\ 0 \\ 17.125 \end{Bmatrix}$$

$= \{311.5,\ 0,\ 4.28\}\ \mu$

$$\begin{Bmatrix} \sigma_x \\ \sigma_y \\ \tau_{xy} \end{Bmatrix}_a = \frac{200(10^3)}{0.91} \begin{bmatrix} 1 & 0.3 & 0 \\ 0.3 & 1 & 0 \\ 0 & 0 & 0.35 \end{bmatrix} \begin{Bmatrix} 311.5 \\ 0 \\ 4.28 \end{Bmatrix}$$

$= \{68.46,\ 20.54,\ 0.33\}$ MPa

8.1

(a) From Eq. (8.13), we have

$$\sigma_{\theta,min}=\frac{a^2p_i}{b^2-a^2}(1+\frac{b^2}{b^2})=p_i\frac{2a^2}{b^2-a^2}$$

$$\sigma_{\theta,max}=\frac{a^2p_i}{b^2-a^2}(1+\frac{b^2}{a^2})=p_i\frac{a^2+b^2}{b^2-a^2}$$

Hence,

$$\frac{\sigma_{\theta,max}}{\sigma_{\theta,min}}=\frac{a^2+b^2}{2a^2}=\frac{a^2+(1.1a)^2}{2a^2}=1.105$$ ◄

(b) Using Eq. (8.16),

$$\sigma_{\theta,max}=-2p_o\,[b^2/(b^2-a^2)]$$

$$\sigma_{\theta,min}=-p_o\,[(a^2+b^2)/(b^2-a^2)]$$

Thus,

$$\frac{\sigma_{\theta,max}}{\sigma_{\theta,min}}=\frac{2b^2}{b^2-a^2}=\frac{2(1.21a)^2}{2.21a^2}=1.1$$ ◄

8.2

Equation (8.20):

$$\sigma_z=\frac{p_ia^2}{b^2-a^2}=\frac{0.6^2p_i}{1^2-0.6^2}=0.5625p_i=140$$

or

$$p_i=248.9\text{ MPa}$$

Equation (8.13):

$$\sigma_{\theta,max}=\frac{b^2+a^2}{b^2-a^2}p_i=\frac{1^2+0.6^2}{1-0.6^2}p_i=2.12p_i=140$$

or

$$p_i=65.9\text{ MPa}$$

Equation (8.10):

$$\tau_{max}=\frac{p_ib^2}{b^2-a^2}=\frac{1^2}{1^2-0.6^2}p_i=1.5625p_i=80$$

or

$$p_i=51.2\text{ MPa}=p_{all}$$ ◄

8.3

(a) Initial maximum tangential stress, from Eq. (8.13),

$$\sigma_\theta=p_i\frac{b^2+a^2}{b^2-a^2}=\frac{n^2+1}{n^2-1}p_i$$

or

$$p_i=\sigma_\theta[(n^2-1)/(n^2+1)]$$

After boring, denoting the inner radius by r_x, we have

(CONT.)

8.3 CONT.

$$\Delta\sigma_\theta+\sigma_\theta=p_i\frac{n^2a^2+r_x^2}{n^2a^2-r_x^2}=\frac{n^2-1}{n^2+1}\sigma_\theta\frac{n^2a^2+r_x^2}{n^2a^2-r_x^2}$$

or

$$(\Delta\sigma_\theta+\sigma_\theta)(n^2+1)(n^2a^2-r_x^2)=(n^2-1)\sigma_\theta(n^2a^2+r_x^2)$$

or

$$(\Delta\sigma_\theta+\sigma_\theta)(n^2+1)n^2a^2-(n^2-1)\sigma_\theta n^2a^2=r_x^2[(\Delta\sigma_\theta+\sigma_\theta)(n^2+1)+(n^2-1)\sigma_\theta]$$

This gives the new inner radius in the form

$$r_x=\left[\frac{2n^2a^2\sigma_\theta+\Delta\sigma_\theta(n^2+1)n^2a^2}{\Delta\sigma_\theta(n^2+1)+2\sigma_\theta n^2}\right]^{1/2}$$ ◄

(b)

$$r_x=\left[\frac{2(4)(0.025)^2\sigma_\theta+0.1\sigma_\theta(5)4(0.025)^2}{0.1\sigma_\theta(5)+2\sigma_\theta(4)}\right]^{1/2}$$

$$=0.02712\text{ m}=27.12\text{ mm}$$ ◄

8.4

Using Eq. (8.13),

$$\sigma_{\theta,max}=p_i\,[(a^2+b^2)/(b^2-a^2)]$$

$$\frac{280}{2}=7\,\frac{(0.6)^2+b^2}{b^2-(0.6)^2}$$

Solving, b=0.6308 m. Therefore,

$$t=b-a=630.8-600=30.8\text{ mm}$$ ◄

8.5

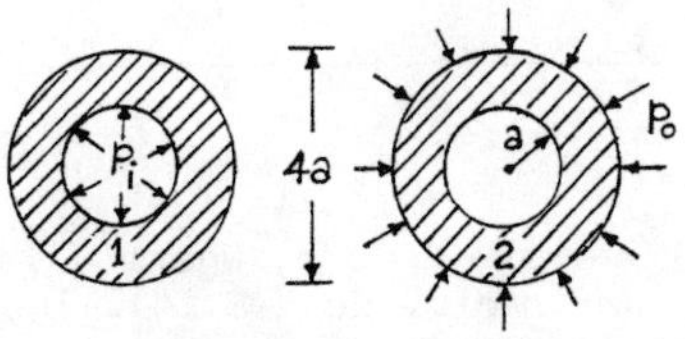

(a) Equations (8.13) and (8.16) give at r=a:

$$\sigma_{\theta1}=p_i\frac{b^2+a^2}{b^2-a^2},\qquad\sigma_{\theta2}=-2p_o\frac{b^2}{b^2-a^2}$$

Then,

$$|\sigma_{\theta1}|=|\sigma_{\theta2}|;\qquad p_i\frac{b^2+a^2}{b^2-a^2}=2p_o\frac{b^2}{b^2-a^2}$$

or

$$\frac{p_i}{p_o}=\frac{2b^2}{a^2+b^2}=\frac{2(4a^2)}{a^2+4a^2}=1.6$$ ◄

(CONT.)

8.5 CONT.

(b) By neglecting the strain ε_L in the longitudinal direction,

$$\varepsilon_{\theta 1}=\frac{1}{E}(\sigma_{\theta 1}+\nu p_i), \quad \varepsilon_{\theta 2}=-\frac{1}{E}(\sigma_{\theta 2}-\nu p_o)$$

Then,

$$|\varepsilon_{\theta 1}| = |\varepsilon_{\theta 2}|$$

gives

$$\sigma_{\theta 1}+\nu p_i = \sigma_{\theta 2}-\nu p_o \qquad \text{(a)}$$

But

$$\frac{\sigma_{\theta 1}}{p_i}=\frac{a^2+b^2}{b^2-a^2}=\frac{a^2+4a^2}{4a^2-a^2}=1.66$$

Hence,

$$\sigma_{\theta 1}=1.66p_i \qquad \text{(b)}$$

We also have

$$\frac{\sigma_{\theta 2}}{p_o}=\frac{2b^2}{b^2-a^2}=\frac{2(4a^2)}{4a^2-a^2}=2.66$$

or

$$\sigma_{\theta 2}=2.66p_o \qquad \text{(c)}$$

Substituting Eqs. (b) and (c) into (c) and letting $\nu=1/3$:

$$\frac{p_i}{p_o}=1.16 \quad ◄$$

8.6

Equation (8.14), substituting the given data yields

$$u=\frac{ap_i}{E}\left(\frac{a^2+b^2}{b^2-a^2}+\nu\right)$$

$$=\frac{0.6(7\times10^6)}{200\times10^9}\left(\frac{0.6^2+0.6308^2}{0.6308^2-0.6^2}+0.3\right)$$

$$=0.426 \text{ mm} \quad ◄$$

8.7

(a) We have $\varepsilon_\theta=u/r$, where u is defined by Eq. (8.14). Thus, at r=a:

$$\varepsilon_{\theta,max}=\frac{p_i}{E}\left[\frac{b^2+a^2}{b^2-a^2}+\nu\right] \qquad \text{(a)}$$

(b) Introducing σ_θ, σ_r, and σ_z from Eqs. (8.12), (8.13) and (8.20) into Hooke's law

$$\varepsilon_\theta=[\sigma_\theta-\nu(\sigma_r+\sigma_z)]/E$$

we have

$$\varepsilon_{\theta,max}=\frac{p_i}{E}\left[\frac{b^2+(1-\nu)a^2}{b^2-a^2}+\nu\right] \qquad \text{(b)}$$

Substituting the data, Eq. (a):

(CONT.)

8.7 CONT.

$$0.001=\frac{60(10^6)}{200(10^9)}\left[\frac{4+a^2}{4-a^2}+\frac{1}{3}\right]$$

Solving, a=1.41 m. Then,

$$t=2-1.41=0.59 \text{ m} = t_{req.} \quad ◄$$

Similarly, Eq. (b) yields

$$0.001=\frac{60(10^6)}{200(10^9)}\left[\frac{4+2a^2/3}{4-a^2}+\frac{1}{3}\right]$$

or

a=1.48 m; t=0.52 m

8.8

Equation (8.13) at r=a:

$$\sigma_{\theta,max}=\frac{0.25^2+0.05^2}{0.25^2-0.05^2}(60)=65 \text{ MPa}$$

Equation (8.12) at r=a:

$$\sigma_{r,max}=-p_i=-60 \text{ MPa}$$

Equation (8.10):

$$\tau_{max}=\frac{60(0.25)^2}{0.25^2-0.05^2}=62.5 \text{ MPa}$$

Equation (8.20) with $p_o=0$:

$$\sigma_z=\frac{0.05^2(60)}{0.25^2-0.05^2}=2.5 \text{ MPa}$$

Stress-strain relationship is given by

$$\varepsilon_\theta=\frac{1}{E}[\sigma_\theta-\nu(\sigma_r+\sigma_z)]=\frac{u}{r} \qquad \text{(a)}$$

Substituting Eqs. (8.12), (8.13), and (8.20), this expression results in at r=a:

$$u=\frac{ap_i}{E(b^2-a^2)}[(1-2\nu)a^2+(1+\nu)b^2] \qquad \text{(P8.8)}$$

Introducing the given numerical values, we obtain

$$u=\frac{0.05(60\times10^6)}{72\times10^6(0.25^2-0.05^2)}\left[0.4(0.05)^2+1.3(0.25)^2\right]$$

$$=0.0571 \text{ mm}$$

The change in the internal diameter is therefore,

$$\Delta d=2u=0.1142 \text{ mm} \quad ◄$$

8.9

Equation (8.19):

$$\sigma_{\theta,max} = -\frac{2(0.25)^2(60)}{0.25^2-0.05^2} = -125 \text{ MPa}$$

Equation (8.15) at r=b:

$$\sigma_{r,max} = -p_o = -60 \text{ MPa}$$

Equation (8.20) for $p_i=0$:

$$\sigma_z = -\frac{0.25^2(60)}{0.25^2-0.05^2} = -62.5 \text{ MPa}$$

Equation (8.9) at r=a and $p_i=0$:

$$\tau_{max} = -\frac{p_o b^2}{b^2-a^2} = -\frac{60(0.25)^2}{0.25^2-0.05^2} = -62.5 \text{ MPa}$$

Substitution of Eqs. (8.15), (8.16), and (8.20), into Eq. (a) of Solution of Prob. 8.8 leads to for r=a:

$$u = -\frac{a p_o b^2}{E(b^2-a^2)}(2-\nu) \qquad \text{(P8.9)}$$

$$= -\frac{0.05(60\times10^6)(0.25^2)}{72\times10^9(0.25^2-0.05^2)}(2-0.3)$$

$$= -0.0738 \text{ mm}$$

Thus,

$$\Delta d = 2u = -0.1476 \text{ mm}$$ ◄

8.10

Given:

b=0.1 m c=0.3 m a=0

$\delta = 0.001(0.1) = 0.0001$ m

Using Eq. (8.23):

$$p = \frac{E\delta}{b}\,\frac{c^2-b^2}{2c^2}$$

$$= \frac{200\times10^9(0.0001)}{0.1}\,\frac{0.09-0.01}{2(0.09)}$$

$$= 88.89 \text{ MPa}$$

Then, letting $p=p_i=88.89$ MPa, Eq. (8.8) gives at r=b:

$$\sigma_\theta = 0 = \frac{b^2p-c^2p_o}{c^2-b^2} + \frac{(p-p_o)c^2}{c^2-b^2}$$

After substituting the numerical values, this equation results in

$$p_o = 49.38 \text{ MPa}$$

8.11

(a) Using Eq. (8.18),

$$\frac{\sigma_{\theta,max}}{p_i} = \frac{a^2+b^2}{b^2-a^2} = \frac{4}{3}$$

or

b=2.65a

Then,

$$b/a = \frac{a+t}{a} = 2.65; \quad t = 1.65a$$

Hence,

$$t/2a = \frac{1.6a}{2a} = 0.825$$ ◄

(b) Neglect longitudinal strain and consider

$\sigma_r = -p_i = 6.3$ MPa

$\sigma_\theta = 4(6.3)/3 = 8.4$ MPa

Therefore, for a=0.075 m and b=2.65a:

$$\Delta d = \varepsilon_d(2a) = \frac{2p_i a}{E}\left[\frac{a^2+b^2}{b^2-a^2}+\nu\right]$$

$$= \frac{2(6.3\times10^6)(0.075)}{210(10^9)}\left[\frac{4}{3}+\frac{1}{3}\right]$$

$$= 7.4(10^{-3}) \text{ mm}$$ ◄

Alternatively, use Eq. (8.14) and let $\Delta d = 2u$.

8.12

Introducing various values of P, as given in Fig. 8.4, into Eqs. (a) and (8.21) of Sec. 8.3, it is seen that σ_θ and S values as shown in the figure are found.

For example, let P=1 or $p_i = p_o$. Then,

$$\sigma_\theta = p_i\frac{1-R^2}{R^2-1} + p_i b^2 \frac{1-1}{(R^2-1)r} \; ; \; \sigma_\theta = -p_i$$ ◄

and

$$S = \sigma_{\theta i}/\sigma_{\theta o} = (1+0)/(1+0) = 1$$

or

$$\sigma_{\theta i} = \sigma_{\theta o} \quad \text{or} \quad S=1$$ ◄

8.13

(a) For $\varepsilon_z=0$; $[\sigma_z - \nu(\sigma_r+\sigma_\theta)]/E = 0$, using Eqs. (8.8),

$$\sigma_z = \nu(\sigma_r+\sigma_\theta) = \frac{2\nu(a^2p_i-b^2p_o)}{b^2-a^2}$$ ◄

(b) Similarly, for $\sigma_z=0$:

$$\varepsilon_z = -\frac{\nu}{E}(\sigma_r+\sigma_\theta) = -\frac{2\nu(a^2p_i-b^2p_o)}{E(b^2-a^2)}$$ ◄

8.14

At r=a: from Eq. (8.18),

$$\sigma_{\theta,max} = \frac{4a^2+a^2}{4a^2-a^2}p_i = \frac{5}{3}p_i$$

and from Eq. (8.12),

$$\sigma_{r,max} = -p_i \,.$$

Energy of distortion theory:

$$p_i\,[(\tfrac{5}{3})^2-(\tfrac{5}{3})(-1)+(-1)^2]^{1/2} = \sigma_{yp}$$

or

$$p_i = 0.429\,\sigma_{yp}$$ ◄

Maximum shearing stress theory:

$$(5p_i/3)-(-p_i) = \sigma_{yp}$$

or

$$p_i = 0.375\,\sigma_{yp}$$ ◄

8.15

We have, at r=a:

$$\sigma_{\theta,max} = \frac{b^2+a^2}{b^2-a^2}p_i = \frac{9a^2+a^2}{9a^2-a^2}p_i = \frac{5}{4}p_i$$

$$\sigma_{r,max} = -p_i$$

(a)

$$|\sigma_u| = |\tfrac{5}{4}p_i|$$

or

$$p_i = 0.8(350) = 280 \text{ MPa}$$ ◄

and

$$|\sigma_u| = |p_i|\,,\quad p_i = 350 \text{ MPa}$$

(b) Using Eq. (4.12a),

$$\frac{5p_i}{4(350)} - \frac{-p_i}{630} = 1$$

Solving,

$$p_i = 193.8 \text{ MPa}$$ ◄

8.16

From Eq. (8.18) for r=a:

$$p_i = 35\,\frac{(0.25)^2-(0.05)^2}{(0.25)^2+(0.05)^2} = 32.31 \text{ MPa}$$

Total radial force on the contact surface is $2\pi aLp$ and total frictional force equals $2\pi apL\mu$. Hence,

$$\text{Torque} = 2\pi apL\mu(a)$$

$$= 2\pi(0.05^3)(32.31\times10^6)(0.2)$$
$$= 5.073 \text{ kN·m}$$ ◄

8.17

Let ε_{s1}, ε_{c1}, and ε_{s2}, ε_{c2} be initial and final compressive and tensile strains in the shaft and in the cylinder, respectively.

Also, let $\sigma_{\theta 1}$, p_1 and $\sigma_{\theta 2}$, p_2 denote the initial and final maximum stresses and contact pressures, respectively. Then,

$$\varepsilon_{s1} + \varepsilon_{c1} = \varepsilon_{s2} + \varepsilon_{c2} \qquad \text{(a)}$$

Here,

$$\varepsilon_{s1} = \frac{1}{E}(p_1 - \nu p_1) = \frac{2}{3E}p_1$$

$$\varepsilon_{c1} = \frac{1}{E}(\sigma_{\theta 1} + \nu p_1) = \frac{p_1}{E}(2+\tfrac{1}{3}) = \frac{7}{3E}p_1$$

$$\varepsilon_{s2} = \frac{1}{E}[(p_2 - \nu p_2) + \frac{\nu P_L}{\pi a^2}]$$

$$= \frac{1}{E}[p_2(1-\tfrac{1}{3}) + \frac{1}{3}\,\frac{4.5}{\pi(0.05)^2}]$$

$$= \frac{1}{E}[\frac{2}{3}p_2 + \frac{1909.86}{3}]$$

From the condition of linearity,

$$\frac{\sigma_{\theta 2}}{\sigma_{\theta 1}} = \frac{p_2}{p_1}\,;\quad \sigma_{\theta 2} = p_2\frac{\sigma_{\theta 1}}{p_1} = p_2\frac{2p_1}{p_1} = 2p_2$$

and

$$\varepsilon_{c2} = \frac{1}{E}(\sigma_{\theta 2} + \nu p_2) = \frac{1}{E}(2p_2 + \nu p_2) = \frac{7p_2}{3E}$$

Equation (a) is thus,

$$\frac{2p_1}{3E} + \frac{7p_1}{3E} = \frac{1}{E}(\frac{2}{3}p_2 + \frac{1909.86}{3}) + \frac{7p_2}{3E}$$

from which

$$p_1 = p_2 + 636.62$$

Hence,

$$\Delta p = p_2 - p_1 = -636.62 \text{ kPa}$$ ◄

8.18

From Eq. (8.22),

$$\delta = \frac{bp}{E_b}\left(\frac{b^2+c^2}{c^2-b^2} + \nu_b\right) + \frac{bp}{E_s}\left(\frac{a^2+b^2}{b^2-a^2} - \nu_s\right)$$

or

$$p = \frac{\delta}{b\left[\frac{1}{E_b}\left(\frac{c^2+b^2}{c^2-b^2} + \nu_b\right) + \frac{1}{E_s}\left(\frac{a^2+b^2}{b^2-a^2} - \nu_s\right)\right]}$$

Substituting the given data, this gives p=9.01 MPa

Stresses in the steel cylinder, using Eq. (8.16),

$$(\sigma_\theta)_{r=0} = -\frac{2(9.01\times10^6)(0.08)^2}{(0.08)^2-(0.04)^2} = -24.03 \text{ MPa}$$

(CONT.)

8.18 CONT.

$$(\sigma_\theta)_{r=0.08} = -9.01(10^6)\frac{(0.08)^2+(0.04)^2}{(0.08)^2-(0.04)^2}$$

$$= -15.02 \text{ MPa}$$ ◄

Stresses in the brass cylinder, from Eq. (8.13):

$$(\sigma_\theta)_{r=0.08} = 9.01(10^6)\frac{(0.14)^2+(0.08)^2}{(0.14)^2-(0.08)^2}$$

$$= 17.75 \text{ MPa}$$ ◄

$$(\sigma_\theta)_{r=0.14} = \frac{2(9.01\times10^6)(0.08)^2}{(0.14)^2-(0.08)^2} = 8.74 \text{ MPa}$$ ◄

8.19

(a) Using Eq. (P8.18),

$$p = \frac{(0.5/2)(10^{-3})}{0.1\left[\frac{1}{72(10^9)}\left(\frac{0.15^2+0.1^2}{0.15^2-0.1^2}+0.33\right)+\frac{1-0.29}{200(10^9)}\right]}$$

$$= 56.49 \text{ MPa}$$ ◄

(b) From Eq. (8.14), for r=b:

$$u = \frac{2a^2bp_i}{E(b^2-a^2)} = \frac{2(0.1)^2(0.15)(56.49\times10^6)}{72\times10^9(0.15^2-0.1^2)}$$

$$= 0.1883 \text{ mm}$$

$$\delta = 2u = 0.3766 \text{ mm}$$ ◄

8.20

Equation (8.14) at r=b:

$$u = \frac{\delta_o}{2} = \frac{bp}{E}\left(\frac{b^2+c^2}{c^2-b^2}+\nu\right) = \frac{bp}{E}\left(\frac{b^2+2.25b^2}{2.25b^2-b^2}+\frac{1}{3}\right)$$

Solving,

$$p = E\delta_o/5.86b$$

Then, Eq. (8.17) yields at r=b:

$$u = \frac{bp}{E}(1-\nu) = \frac{2\delta_o}{3(5.86)} = 0.114\delta_o$$

Hence,

$$\Delta d = 2u = 0.23\,\delta_o$$ ◄

8.21

(a) Initial difference in diameter:

$$\Delta = 2b(\varepsilon_1+\varepsilon_2)$$

where,

ε_1 = tangential (comp.) strain in the shaft

(CONT.)

8.21 CONT.

ε_2 = tangential (tens.) strain in the cylinder

Thus,

$$\Delta = \frac{2b}{E}[(p-\nu p)+(\sigma_{\theta,max}+\nu p)]$$

$$= \frac{2b}{E}[(\sigma_{\theta,max}+p) = \frac{2b}{E}(2p+p)$$

$$= 6pb/E$$

(b) Compressive (uniform) strain ε_L, due to axial load P, is

$$\varepsilon_L = P/\pi b^2E$$

We now have

$$\varepsilon_1 = \frac{1}{E}(p_1 - \nu p_1 - \frac{\nu P}{\pi b^2}) \qquad \text{(comp.)}$$

$$\varepsilon_2 = \frac{1}{E}(\sigma_{\theta 1,max} + \nu p_1) \qquad \text{(tens.)}$$

where $\sigma_{\theta 1,max}$ is the increased tangential stress. Thus,

$$\Delta_1 = 2b(\varepsilon_1+\varepsilon_2) = \frac{2b}{E}(\sigma_{\theta 1,max}+p_1-\frac{\nu P}{\pi b^2})$$

Based on the linearity condition:

$$\frac{\sigma_{\theta 1,max}}{p_1} = \frac{\sigma_{\theta,max}}{p} = 2$$

or

$$\sigma_{\theta 1,max} = 2p_1$$

Setting $\Delta = \Delta_1$:

$$\frac{6pb}{E} = \frac{2b}{E}(2p_1+p_1-\frac{P}{3\pi b^2})$$

or

$$P = 9\pi b^2(p_1-p)$$ ◄

8.22

Radial strain is

$$\varepsilon = \varepsilon_{b,comp.} + \varepsilon_{s,tens.}$$

$$= \frac{1}{E_b}(p-\nu p)+\frac{1}{E_s}(\sigma_{\theta,max}+\nu p)$$

We also have

$$\varepsilon = (T_2-T_1)(\alpha_b-\alpha_s)$$

$$= \Delta T(19.5-11.7)10^{-6} = 7.8(10^{-6})\Delta T$$

Note that from Eq. (8.18) at r=a:

$$\frac{\sigma_{\theta,max}}{p} = \frac{4b^2+b^2}{4b^2-b^2} = \frac{5}{3}$$

Thus,

$$7.8(10^{-6})\Delta T = \frac{1}{E_b}(p-\frac{p}{3})+\frac{1}{E_s}(\frac{5p}{3}+\frac{p}{3})$$

or

$$p = \frac{11.7(\Delta T)E_bE_s}{(E_s+3E_b)10^6}$$

Hence,

$$\sigma_{\theta,max} = \frac{1.95E_bE_s(T_2-T_1)}{(E_s+3E_b)10^5}$$ ◄

8.23

We have (Fig. 8.5):

a=0.05 m   b=0.1 m   c=0.15 m

Maximum tangential stress occurs at r=0.1 m in gear wheel:

$$\sigma_{max}=p\,\frac{c^2+b^2}{c^2-b^2}$$

where, p is the internal pressure exerted by the contact surfaces. Substituting the given data into this equation, we have

$$0.21=p\,\frac{0.15^2+0.1^2}{0.15^2-0.1^2}$$

Solving,

$$p=0.081\ MPa=81\ kPa$$

This interface pressure produces the torque at the contact surface. Area of contact is

$$2\pi bL=2\pi(0.1)(0.1)=0.02\pi$$

The torque transmitted is thus,

$$T=[(0.2)(81\times10^3)(0.02\pi)](0.1)$$

$$=101.79\ N\cdot m \qquad \blacktriangleleft$$

8.24

From the first of Eqs. (8.28),

$$\frac{\partial\sigma_r}{\partial r}=\frac{3+\nu}{8}\,\rho\omega^2[b^2+a^2-\frac{\partial}{\partial r}(r^2)-a^2b^2\frac{\partial}{\partial r}(\frac{1}{r^2})]=0$$

from which

$$\frac{2a^2b^2}{r^3}-2r=0; \qquad r=\sqrt{ab} \qquad \blacktriangleleft$$

This value of r is substituted into the first of Eqs. (8.28) to yield

$$\sigma_{r,max}=\rho\omega^2(b-a)^2 \qquad \blacktriangleleft$$

We also find from the second of Eqs. (8.28), for r=a:

$$\sigma_{\theta,max}=2\,\frac{3+\nu}{8}\,\rho\omega^2\left(b^2+\frac{1-\nu}{3+\nu}\,a^2\right)$$

Thus,

$$\frac{\sigma_{\theta,max}}{\sigma_{r,max}}=\frac{2\left(b^2+\frac{1-\nu}{3+\nu}\,a^2\right)}{(b-a)^2} \qquad \blacktriangleleft$$

8.25

At r=0: $\sigma_\theta=\sigma_r=(3+\nu)b^2\rho\omega^2/8$. Thus,

$$\sigma_\theta^2-\overset{0}{\sigma_r}\sigma_\theta+\overset{0}{\sigma_r^2}=\sigma_{yp}^2$$

or

$$\omega_{all}=\frac{1}{b}\sqrt{\frac{8\sigma_{yp}}{(3+\nu)\rho}} \qquad (P8.25)$$

$$=\frac{1}{0.125}\left[\frac{8(260\times10^6)}{10(2.7\times10^3)/3}\right]^{1/2}$$

$$=3845.9\ rad/sec$$

$$=36,726\ rpm \qquad \blacktriangleleft$$

8.26

(a) When p=0, at interface,

$$u_d-u_s=0.05(10^{-3}) \qquad (a)$$

Use Eq. (8.28) with r=a:

$$(\sigma_\theta)_d=\frac{\rho\omega^2}{4}[(3+\nu)b^2+(1-\nu)a^2]$$

$$=\frac{7.8(10^3)\omega^2}{4}[3.3(0.125^2)+$$

$$0.7(0.025^2)]=101.473\,\omega^2$$

Radial displacement of disk (with $\sigma_r=0$):

$$(u_d)_{r=a}=\frac{(\sigma_\theta)_d\,a}{E}=\frac{101.473\omega^2a}{200(10^9)}$$

For shaft, using Eq. (8.29) with r=a:

$$(\sigma_\theta)_s=\frac{3.3(0.025)^2}{8}(1-\frac{1.9}{3.3})7.8(10^3)\omega^2$$

$$=0.8531\,\omega^2$$

Radial displacement of shaft (with $\sigma_r=0$):

$$(u_s)_{r=a}=\frac{(\sigma_\theta)_s\,a}{E}=\frac{0.8531\omega^2a}{200(10^9)}$$

Thus, Eq. (a) give

$$\frac{101.473\omega^2(0.025)}{200(10^6)}-\frac{0.8531\omega^2(0.025)}{200(10^6)}=0.05$$

or

$$\omega=1,993.884\ rad/sec$$

$$=19,040\ rpm \qquad \blacktriangleleft$$

(b) Maximum stress occurs in disk

$$(\sigma_\theta)_{max}=101.473(1,993.884)^2$$

$$=403.4\ MPa \qquad \blacktriangleleft$$

8.27

We have

$a=0.5c \quad b=2c \quad r=\sqrt{ab}=c$

(a) Equation (8.28):

$$50(10^6)=\frac{3.3c^2}{8}(0.25+4-1-\frac{1}{1}) \times 7.8(10^3)(5000\times\frac{2\pi}{60})^2$$

or

$c=0.1587$ m

Thus,

$t_r=1.5c=238.1$ mm ◄

(b) Equation (8.28) at r=a:

$$\sigma_{\theta,max}=\frac{3.3c^2}{8}[0.25+4-\frac{1.9}{3.3}(0.25)+4] \times 7.8(10^3)(5000\times\frac{2\pi}{60})^2$$

$=180.1$ MPa ◄

8.28

(a) Equation (8.23), with a=0:

$$p=\frac{210(10^9)(0.000075)}{2(0.075)}\,\frac{0.16-0.005625}{0.16}$$

$=101.31$ MPa ◄

Then, Eq. (8.18) gives

$$\sigma_{\theta,max}=101.31(10^6)\,\frac{0.16+0.005625}{0.16-0.005625}$$

$=108.68$ MPa ◄

(b) Applying Eq. (8.28),

$$0.000075=\frac{3.3(0.7)}{8(210\times10^9)}[0.005625+0.16-\frac{1.3}{3.3}(0.005625)+\frac{1.3}{0.7}(0.16)] \times (7.8\times10^3)(0.075)\omega^2$$

Solving,

$\omega=450$ rad/sec

$=4300$ rpm ◄

8.29

From Eqs. (8.29), for r=0 and $\nu=1/3$:

$$\sigma_{max}=\frac{3+\nu}{8}\rho(\omega b)^2$$

$$=\frac{5}{12}\rho v^2$$ ◄

8.30

We have $\frac{t_i}{t_o}=(\frac{b}{a})^s$; $\quad \frac{0.125}{0.0625}=(\frac{0.625}{0.125})^s$

Solving, s=0.431. Then,

$$m_{1,2}=-\frac{0.431}{2}\pm[(\frac{0.431}{2})^2+(1+0.3\times0.431)]^{1/2}$$

or

$m_1=0.869 \quad m_2=-1.3$

Equation (8.33) gives

$$\sigma_r=\frac{c_1}{t_1}r^{0.3}+\frac{c_2}{t_1}r^{-1.869}-0.50169\rho(\omega r)^2 \quad (a)$$

Given conditions are:

$(\sigma_r)_{r=0.625}=0, \quad (\sigma_r)_{r=0.125}=0$

Thus, substituting the given data into Eq. (a) and solving:

$\frac{c_1}{t_1}=0.2322\rho\omega^2, \quad \frac{c_2}{t_1}=-0.0024\rho\omega^2$

The second of Eqs. (8.33), at the bore r=0.125 m, leads to

$$140(10^6)=0.2322(0.125^{0.3})(0.869)\rho\omega^2 +(-0.0024)(-1.3)(0.125^{-1.869})\rho\omega^2 -\frac{(1+0.9)(0.125)^2}{8-(3.3)(0.431)}\rho\omega^2$$

or

$\rho\omega^2=549(10^6)$

It follows, from the second of Eqs. (8.33), that

$$(\sigma_\theta)_{r=0.625}=[0.2322(0.625^{0.3})(0.869) -0.0024(-1.3)(0.625^{-1.869}) -\frac{1.9(0.625)^2}{8-3.3(0.431)}](549\times10^6)$$

$=37.5$ MPa

Circumferential force is thus

37,500(0.0625)=2344 kN/m ◄

8.31

(a) Uniform thickness

Centrifugal force due to the blades, $m\omega^2 r$, is

$$\frac{540}{9.81}(\frac{10{,}000\pi}{60})^2(0.575)=8.677(10^6)\text{ N}$$

(CONT.)

8.31 CONT.

The pressure at b is then

$$p_o=\frac{8.667(10^6)}{2\pi(0.5)(0.05)}=55.24 \text{ MPa}$$

We have
$$\rho\omega^2=7.8(10^{-3})\left(\frac{10,000\pi}{60}\right)^2=2.1384(10^3)$$

The condition of zero pressure at ther bore is satisfied by:

$$(\sigma_r)_{r=a}=0=\frac{E}{1-\nu^2}\left[-\frac{(3+\nu)(1-\nu^2)\rho\omega^2a^2}{8E}+(1+\nu)c_1-(1-\nu)\frac{c_2}{a^2}\right]$$

or

$$0=-\frac{(3+\nu)a^2\rho\omega^2}{8}+\underbrace{\left[\frac{E(1+\nu)}{1-\nu^2}c_1\right]}_{A_1}+\underbrace{\left[-\frac{E(1-\nu)}{1-\nu^2}c_2\right]}_{A_2}\frac{1}{a^2}$$

$$0=A_1+A_2\frac{1}{a^2}-(3+\nu)a^2\rho\omega^2/8$$

Substituting the data, we have

$$0=A_1+256A_2-3.4456 \qquad (a)$$

The condition at the outer circumference is satisfied by:

$$(\sigma_r)_{r=b}=55.24=A_1+A_2/(0.5)^2-3.3(0.5)^2(2138.4)/8$$

or

$$55.24=A_1+4A_2-220.52 \qquad (b)$$

Solution of Eqs. (a) and (b) gives

$$A_1=280.08 \qquad A_2=-1.08 \qquad (c)$$

The second of Eqs. (8.27) is also written as follows

$$\sigma_\theta=A_1-A_2\frac{1}{r^2}-(1+3\nu)\rho\omega^2r^2/8 \qquad (d)$$

where A_1 and A_2 are given by (c). At r=a, we have from Eq. (d):

$$(\sigma_\theta)_{r=a}=280.08+276.48-1.98$$

$$=554.58 \text{ MPa}=\sigma_{max}$$ ◄

Similarly, Eq. (d) gives at r=b:

$$(\sigma_\theta)_{r=b}=280.08+4.32-126.97=157.43 \text{ MPa}$$

Note that $\sigma_{r,max}$ occurs at $r=\sqrt{ab}=0.1768$ m. Thus, Eq. (8.27):

$$(\sigma_r)_{r=0.1768}=-27.67+280-34.55$$

$$=217.88 \text{ MPa}$$

(CONT.)

8.31 CONT.

(b) **Hyperbolic section**

We have

$$\frac{0.4}{0.05}=\left(\frac{0.5}{0.0625}\right)^s; \qquad s=1$$

Then,

$$m_{1,2}=-\frac{1}{2}\pm[0.25+1.3]^{1/2}$$

or

$$m_1=0.745 \qquad m_2=1.745$$

Letting

$$c_1/t_1=B_1 \qquad c_2/t_1=B_2$$

Eq. (8.33) becomes then

$$\sigma_r=B_1r^{m_1+s-1}+B_2r^{m_2+s-1}-\frac{(3+\nu)\rho\omega^2}{8-(3+\nu)s}r^2$$

and

$$(\sigma_r)_{r=a}=0$$
$$=B_1(0.0625)^{0.745}+B_2(0.0625)^{-1.745}-0.7021(0.0625)^2\rho\omega^2$$

or

$$0=B_1+996.35B_2-0.0216\rho\omega^2 \qquad (e)$$

Also

$$(\sigma_r)_{r=b}=55.24-B_1(0.5)^{0.745}+B_2(0.5)^{-1.745}-0.702(0.5)^2\rho\omega^2$$

or

$$55.24=0.55967B_1+3.352B_2-0.1755\rho\omega^2 \qquad (f)$$

Substituting the given data and solving Eqs. (e) and (f):

$$B_1=725.4 \qquad B_2=-0.6814 \qquad (g)$$

The second of Eqs. (8.33),

$$\sigma_\theta=B_1m_1r^{m_1+s-1}+B_2m_2r^{m_2+s-1}-\frac{(1+3\nu)\rho\omega^2}{8-(3+\nu)s}r^2$$

gives at r=a:

$$(\sigma_\theta)_{r=a}=$$

$$725.14(0.745)(0.0625)^{0.745}-0.6814(-1.745)(0.0625)^{-1.745}$$

$$-0.4043(0.0625)^2\rho\omega^2$$

or

$$(\sigma_\theta)_{r=a}=215.2 \text{ MPa}=\sigma_{max}$$ ◄

Also similarly we obtain that

$$(\sigma_\theta)_{r=b}=110.2 \text{ MPa}$$

CONT.)

8.31 CONT.

(c) **Uniform stress**

Using Eq. (8.35),

$$\frac{t_o}{t_1} = e^{-(\rho\omega^2/2\sigma)b^2}$$

where,

$$-(\rho\omega^2/2\sigma)b^2 = \frac{-7.8\times10^3(523.59)^2(0.5)^2}{2(84)}$$

$$=-3.182$$

Thus,

$$t_o = 0.02425e^{-3.182}$$
$$=0.001 \text{ m}=1 \text{ mm}$$

and

$$t=0.02425e^{-12.728r^2}$$ ◄

8.32

We substitute the given T into Eqs. (8.38) to obtain

$$\sigma_r = \frac{\alpha E(T_1-T_2)}{2\ln(b/a)}[-\ln\frac{b}{a} - \frac{a^2(r^2-b^2)}{r^2(b^2-a^2)}\ln\frac{b}{a}]$$

Then,

$$\frac{\partial\sigma_r}{\partial r} = 0 = \frac{E\alpha(T_1-T_2)}{2\ln(b/a)}\left\{\frac{d}{dr}(\ln\frac{b}{a}) - \frac{a^2}{b^2-a^2}[1-\frac{d}{dr}(\frac{b^2}{r^2})]\ln\frac{b}{a}\right\}$$

yields, after differentiation:

$$r=ab\left(\frac{2}{b^2-a^2}\ln\frac{b}{a}\right)^{1/2}$$ ◄

It it noted that, Eq. (8.47) gives the same result.

8.33

Using Eq. P8.32, we have

$$r=0.01(0.015)[\frac{2\ln(1.5)}{(0.015)^2-(0.01)^2}]^{1/2}$$

$$=12.1 \text{ mm}$$

Equations (8.47) are therefore

$$(\sigma_r)_{r=12.1} = \frac{10.4(10^{-6})(90\times10^9)(-8)}{2(0.7)\ln(1.5)} \times$$

$$[-\ln\frac{15}{12.1} - \frac{100(12.1^2-15^2)}{12.1^2(15^2-10^2)}\ln(1.5)]$$

$$=1.319(10^7)[0.041]$$

(CONT.)

8.33 CONT.

$$(\sigma_r)_{r=12.1} = 0.541 \text{ MPa}$$

$$(\sigma_r)_{r=10} = 1.319(10^7)[1-\ln(1.5) - \frac{100(10^2+15^2)}{100(15^2-10^2)}\ln(1.5)]$$

$$=6.042 \text{ MPa}$$

$$(\sigma_\theta)_{r=15} = 1.319(10^7)[1-\ln(1) - \frac{100(15^2+15^2)}{100(15^2-10^2)}\ln(1.5)]$$

$$=-4.64 \text{ MPa}$$

$$(\sigma_\theta)_{r=12.1} = 1.319(10^7)[1-\ln(\frac{15}{12.1}) - \frac{100(12.1^2+15^2)}{12.5^2(15^2-10^2)}\ln(1.5)]$$

$$=0.488 \text{ MPa}$$

Similarly,

$$(\sigma_z)_{r=10} = -1.319(10^7)[1-2\ln(1.5) - \frac{2(10^2)}{15^2-10^2}\ln(1.5)]$$

$$=6.055 \text{ MPa} = \sigma_{max}$$ ◄

$$(\sigma_z)_{r=15} = -1.319(10^7)[1-2\ln(1) - \frac{2(10^2)}{15^2-10^2}\ln(1.5)]$$

$$=-4.64 \text{ MPa}$$

8.34

Introducing

$$T(r)=T_o(b-r)/b$$

into Eqs. (8.39), after integration, we obtain the following expressions for stresses

$$\sigma_r = \frac{1}{3}T_o(\frac{r}{b}-1)\alpha E$$

$$\sigma_\theta = \frac{1}{3}T_o(\frac{2r}{b}-1)\alpha E$$ ◄

From these we observe that, at r=b: the radial stress vanishes while tangential stress assumes its maximum value.

8.35

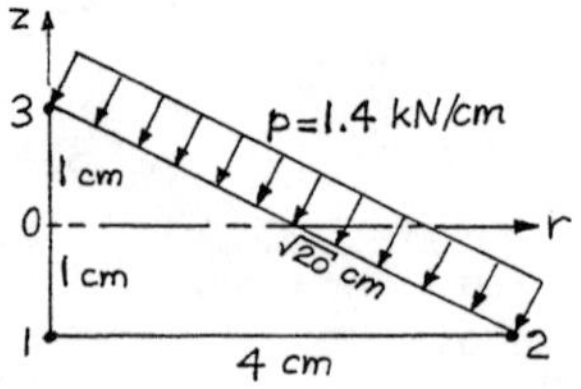

Equivalent nodal force matrix

We have

$r_1=0 \quad z_1=-1 \quad \bar{r}=4/3$

$r_2=4 \quad z_2=-1 \quad \bar{z}=-1/3$

$r_3=0 \quad z_3=1$

Weight of the element is

$$2\pi\bar{r}A(0.077)=2\pi(4/3)(4)(0.077)=2.58 \text{ N}$$

Body forces:

$$\{Q\}_e=\{Q_{r1},Q_{r2},Q_{r3},Q_{z1},Q_{z2},Q_{z3}\}$$

Therefore,

$$\{Q\}_e^b=\{0,0,0,-0.86,-0.86,-0.86\} \text{ N}$$

Surface forces:

$$q_2=q_3=\frac{1400\sqrt{20}}{2}=3130.4 \text{ N/cm}$$

$$q_{r2}=q_{r3}=-3130.4\left(\frac{2}{\sqrt{20}}\right)=-1400 \text{ N/cm}$$

$$q_{z2}=q_{z3}=-3130.4\left(\frac{4}{\sqrt{20}}\right)=-2800 \text{ N/cm}$$

Since $\quad Q_r=2\pi\bar{r}q_r \quad Q_z=2\pi\bar{r}q_z$

Then,

$Q_{r2}=Q_{r3}=-11.729$ kN

$Q_{z2}=Q_{z3}=-23.457$ kN

Therefore,

$$\{Q\}_e^p=\{0,-11.729,-11.729,0,-23.457,-23.457\} \text{ kN}$$

Thermal forces:

$$\{Q\}_e^t=2\pi\bar{r}A[\bar{B}]^T[D][B]\{\varepsilon_0\}$$

Here,

$$[D]=38.46(10^6)\begin{bmatrix}0.3 & 0.7 & 0.3 & 0\\ 0.7 & 0.3 & 0.3 & 0\\ 0.3 & 0.3 & 0.7 & 0\\ 0 & 0 & 0 & 0.2\end{bmatrix}$$

(CONT.)

8.35 CONT.

We also have

$a_i=a_1=4 \quad b_1=-2$

$a_j=a_2=0 \quad b_2=2$

$a_m=a_3=4 \quad b_3=0$

$c_1=-4 \quad d_1=\frac{a_1}{\bar{r}}+b_1+\frac{c_1\bar{z}}{\bar{r}}=2$

$c_2=0 \quad d_2=2$

$c_3=4 \quad d_3=2$

Therefore,

$$[B]=\frac{1}{8}\begin{bmatrix}-2 & 2 & 0 & 0 & 0 & 0\\ 0 & 0 & 0 & -4 & 0 & 4\\ 2 & 2 & 2 & 0 & 0 & 0\\ -4 & 0 & 4 & -2 & 2 & 0\end{bmatrix}$$

It follows that

$$\{Q\}_e^t=2\pi\left(\frac{4}{3}\right)4[\bar{B}]^T[D][\bar{B}]\begin{Bmatrix}0.0006\\0.0006\\0.0006\\0\end{Bmatrix}$$

$$=\{0,502.3,251.2,-502.3,0,502.3\} \text{ kN}$$

Equivalent nodal force matrix is thus,

$$\{Q\}_e=\{Q\}_e^b+\{Q\}_e^p+\{Q\}_e^t$$

where $\{Q\}_e^b$, $\{Q\}_e^p$, $\{Q\}_e^t$ are already obtained.

Element stiffness matrix

$$[k]_e=2\pi\bar{r}A[\bar{B}]^T[D][\bar{B}]$$

$$=2\pi\left(\frac{4}{3}\right)4[\bar{B}]^T[D][\bar{B}]$$

Substituting the values of [D] and $[\bar{B}]$, after carrying out the matrix multiplications, we obtain:

$$[k]_e=\begin{bmatrix}4.8 & 1.6 & -1.6 & 4.8 & -1.6 & -3.2\\ 1.6 & 6.4 & 4 & -8 & 0 & 8\\ -1.6 & 4 & 6 & -4 & 1.6 & 2.4\\ 4.8 & -8 & -4 & 5.6 & -0.8 & -4.8\\ -1.6 & 0 & 1.6 & -0.8 & 0.8 & 0\\ -3.2 & 8 & 2.4 & -4.8 & 0 & 4.8\end{bmatrix}$$

$\times\ (20.14\times10^6)$

Element governing equation

$$\{Q\}_e=[k]_e\{\delta\}_e \qquad \text{(a)}$$

Here,

$$\{\delta\}_e=\{u_1,u_2,u_3,w_1,w_2,w_3\}$$

When boundary conditions are given, $\{\delta\}_e$ is computed from Eq. (a), as illustrated in Chap. 7.

9.1

Using Eq. (9.3),

$$\beta = \left(\frac{k}{4EI}\right)^{1/4} = \left[\frac{1.4\,(10^6)}{4(200\times 10^9)(5.04\times 10^{-6})}\right]^{1/4}$$

$$=0.7676\ \text{m}^{-1}$$

Equation (9.8) yields, for x=0:

$$M_{max} = \frac{-P}{4\beta} f_3(\beta x) = -\frac{P}{4\beta} f_3(0) = -\frac{P}{4\beta}$$

Thus,

$$P = -\sigma_{max}\, I(4\beta)/c$$

$$= \frac{210\times 10^6 (5.04\times 10^{-6})\, 4(0.7676)}{0.0635}$$

$$=51.18\ \text{kN}$$ ◄

9.2

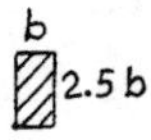

$$I = b(2.5b)^3/12$$
$$=1.302b^4$$

$$M_{max} = \frac{\sigma_{max} I}{c} = \frac{250\times 10^6 (1.302 b^4)}{1.25 b}$$

$$=260.4(10^6)b^3 \qquad \text{(a)}$$

$$\beta = \left[\frac{20\,(10^6)}{4(200\times 10^9)(1.302 b^4)}\right]^{1/4}$$

$$=0.0662/b$$

Equation (9.8):

$$M_{max} = \frac{P}{4\beta} \qquad \text{(b)}$$

From Eqs. (a) and (b),

$$260.4(10^6)b^3 = \frac{40(10^3)\,b}{4(0.0662)}$$

or

$$b=0.0241\ \text{m}=24.1\ \text{mm}$$ ◄

9.3

Select a particular solution of the form

$$\vartheta_p = a \sin\frac{2\pi x}{L}, \qquad a=\text{const.}$$

Introduce this into Eq. (9.1),

$$\left(\frac{2\pi}{L}\right)^4 a + \frac{k}{EI} a = \frac{p_1}{EI}$$

or

$$a = \frac{p_1}{k + \frac{16\pi^4 EI}{L^4}} = \frac{p_1}{k[1+4(\pi/\beta L)^4]}$$

Thus,

$$\vartheta = \frac{p_1}{k + \frac{16\pi^4 EI}{L^4}} \sin\frac{2\pi x}{L} \qquad \text{(a)}$$

(CONT.)

9.3 CONT.

General solution is

$$\vartheta = e^{\beta x}[A\cos\beta x + B\sin\beta x] + e^{-\beta x}[C\cos\beta x + D\sin\beta x] + \vartheta_p$$

Boundary conditions are:

$$\vartheta(\infty)=0; \qquad A=B=0$$

Loading repeats itself periodically, $p_1 \sin(2\pi x/L)$. But

$$e^{-\beta x}[C\cos\beta x + D\sin\beta x]$$

cannot repeat periodically and represents a damped wave. Thus, in order deflection to repeat itself with the same wave length as loading it is required that C=D=0. Accordingly, solution is

$$\vartheta = \vartheta_c + \vartheta_p = 0 + \vartheta_p = \vartheta_p$$

given by Eq. (a).

9.4

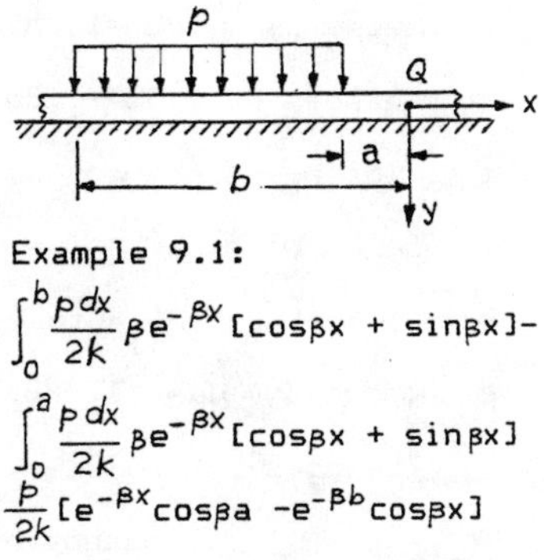

From Example 9.1:

$$\vartheta_Q = \int_0^b \frac{p\,dx}{2k}\beta e^{-\beta x}[\cos\beta x + \sin\beta x] - \int_0^a \frac{p\,dx}{2k}\beta e^{-\beta x}[\cos\beta x + \sin\beta x]$$

$$= \frac{p}{2k}[e^{-\beta x}\cos\beta a - e^{-\beta b}\cos\beta x]$$

Note that, if a=0 and b=L (large):

$$\vartheta_Q \approx \frac{p}{2k}$$

When a and b increase (large):

$$\vartheta_Q \to 0 \qquad \text{(a)}$$

While according to the <u>result of Example 9.1</u>, when a=0 and b=L (large):

$$\vartheta_Q \approx \frac{p}{2k}$$

When a and b increase (large);

$$\vartheta_Q \approx \frac{p}{2k} \qquad \text{(b)}$$

the answers differ, as observed by comparing Eqs. (a) and (b).

9.5

Applying Eqs. (9.3) and (9.8) we obtain

$$\beta = \left[\frac{k}{4EI}\right]^{1/4} = \left[\frac{16.8}{4(8.437)}\right]^{1/4} = 0.84 \text{ m}^{-1}$$

$$\upsilon(0) = \upsilon_{max} = P\beta/2k$$

$$= \frac{0.135(0.84)}{2(16.8)} = 3.375(10^{-3}) \text{ m} \quad ◄$$

and

$$M(0) = M_{max} = \frac{P}{4\beta} = \frac{135}{4(0.84)} = 40.179 \text{ kN·m}$$

Maximum stress is therefore

$$\sigma_{max} = \frac{M_{max}}{S} = \frac{40.179(10^3)}{3.9(10^{-4})} = 103.02 \text{ MPa} \quad ◄$$

9.6

Equations (9.3) and (9.8):

$$\beta = \left[\frac{k}{4EI}\right]^{1/4}, \quad \upsilon_{max} = \frac{P\beta}{2k}, \quad M_{max} = \frac{P}{4\beta}$$

(a)

k is 1.25 times the actual value.

β changes by $\sqrt[4]{1.25} = 1.057$.

υ_{max} changes by 1.057/1.25=0.846.

σ_{max} (or M_{max}) changes by 1/1.057=0.946.

(b)

k is 1.4 times the actual value.

β changes by $\sqrt[4]{1.4} = 1.088$.

υ_{max} changes by 1.088/1.4=0.777.

σ_{max} (or M_{max}) changes by 1/1.088=0.919.

Note that stress calculation is not affected appreciably in both cases.

9.7

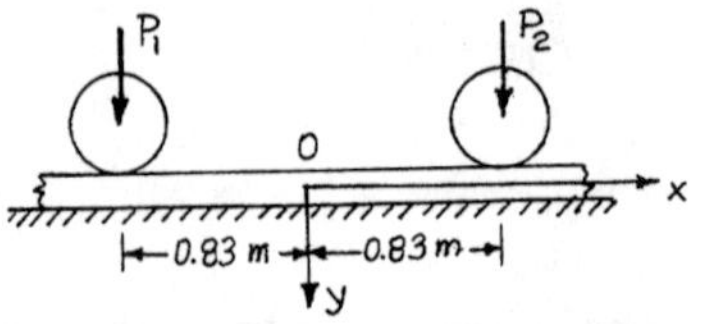

(CONT.)

9.7 CONT.

By using the principle of superposition, deflection of any point, say O, of rail is expressed as the algebraic sum of υ_1 and υ_2 caused by P_1 and P_2, respectively. Thus, from Eq. (9.8), we have

$$\upsilon_0 = \frac{P_1\beta}{2k} f_1(\beta x_1) + \frac{P_2\beta}{2k} f_1(\beta x_2) \quad \text{(a)}$$

where, $P_1 = P_2 = P$, $\beta = 0.84 \text{ m}^{-1}$. Then, $f_1(\beta x_1)$ and $f_1(\beta x_2)$ are found from Table 9.1 for $x_1 = 0.83$ m and $x_2 = -0.83$ m. We have $\beta x_1 = 0.697$ and $\beta x_2 = -0.697$. Thus,

$$f_1(\beta x_1) = 0.702$$

Since, from symmetry $f_1(\beta x_1)$ has the same value for a (+) or (−) value of βx, Eq. (a) may be written as

$$\upsilon_0 = \frac{P\beta}{2k}(0.702 + 0.702) = 1.404\frac{P\beta}{2k} \quad ◄$$

Resultant bending moment at O, from Eq. (9.8), is

$$M_0 = \frac{P}{4\beta} f_3(\beta x_1) + \frac{P}{4\beta} f_3(\beta x_2)$$

$$= \frac{P}{4\beta}[2f_3(0.697)] = 0.125\frac{P}{4\beta} \quad ◄$$

It may be verified by comparing the results of Problems 9.5 and 9.7 that addition of one or more load (reduces appreciably value of maximum moment) causes a large increase in the maximum deflection of the rail.

9.8

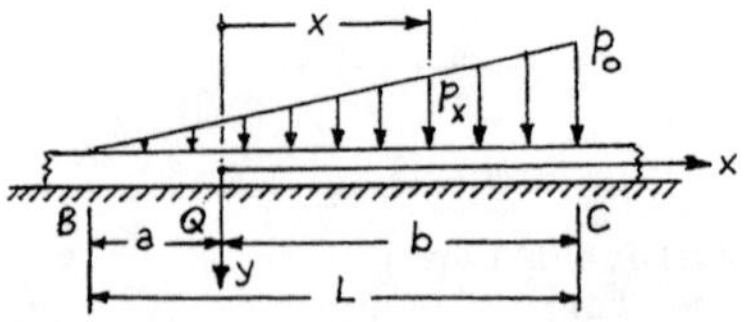

Expressions for loading are:

$$p_x = \frac{p_o}{L}(a - x) \qquad \text{(segment AB)}$$

$$p_x = \frac{p_o}{L}(a + x) \qquad \text{(segment AC)}$$

Deflection at Q is is obtained by substituting ($p_x dx$) for P in Eq. (9.6). That is

(CONT.)

9.8 CONT.

$$v_Q=\frac{P_o\beta}{2kL}\{\int_0^a (a-x)e^{-\beta x}[\cos\beta x+\sin\beta x]dx$$

$$+\int_0^b (a+x)e^{\beta x}[\cos\beta x+\sin\beta x]\}dx$$

Integrating, we have

$$v_Q=\frac{P_o}{4\beta k}[f_3(\beta a)-f_3(\beta b)-2\beta L f_4(\beta b)+4\beta a]$$

9.9

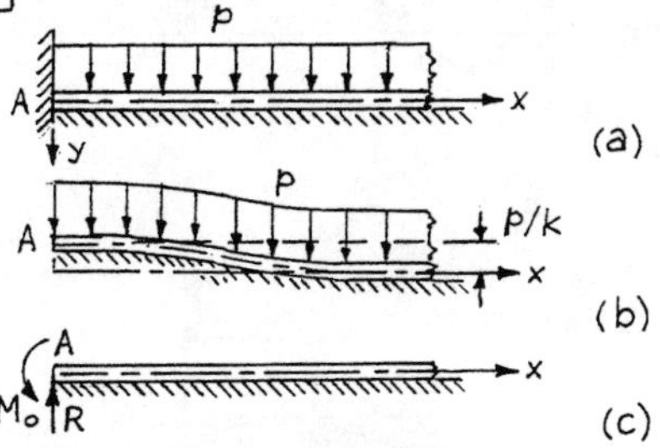

Referring to Fig. (b):

$$v_{A1}=-p/k \qquad \theta_{A1}=0$$

Referring to Fig. (c) and using Eq. (9.12),

$$v_{A2}=\frac{2\beta}{k}[-Rf_4(0)+\beta M_o f_3(0)]$$

$$\theta_{A2}=-\frac{2\beta^2}{k}[-Rf_1(0)+2\beta M_o f_4(0)]$$

The problem may be separated into two different cases both semi-infinite beams (as is seen in the figure above) provided that deflection at A is zero. Thus,

$$v_{A1}-v_{A2}=0; \qquad -\frac{p}{k}=\frac{2\beta}{k}[-R+\beta M_o]$$

$$\theta_{A1}-\theta_{A2}=0; \qquad \frac{2\beta^2}{k}[-R+2\beta M_o]$$

Solving,

$$M_o=2\beta^2 EI\frac{p}{k}, \quad R=4\beta^2 EI\frac{p}{k}=\frac{p}{\beta}$$ ◄

9.10

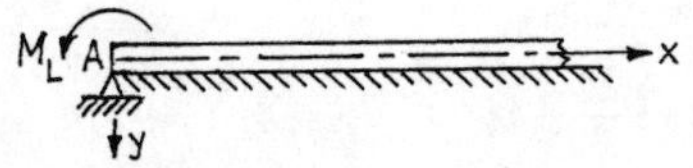

Equation (9.11) with $v(0)=0$ gives

$$0=2\beta(-R_A+\beta M_L)/k$$

from which $R_A=M_L$. When R_A is substituted for $-P$ into Eq. (9.12):

$$v=-\frac{M_L}{2\beta^2 EI}e^{-\beta}\sin\beta x=-\frac{M_L}{2\beta^2 EI}f_2(\beta x)$$

(CONT.)

9.10 CONT.

Successive differentiations of this expression give:

$$v'=-\frac{M_L}{2\beta EI}f_3(\beta x)$$

$$M=\frac{M_L}{2EI}f_4(\beta x)$$ ◄

$$V=-\frac{M_L}{2EI}f_1(\beta x)$$

9.11

We have

$$k=\frac{K_s}{a}=\frac{180}{0.625}=288 \text{ kPa}$$

$$=\left[\frac{288(10^3)}{4(200\times10^9)(5.04\times10^{-6})}\right]^{1/4}$$

$$=0.51697 \text{ m}^{-1}$$

Using Eq. (9.8),

$$v(0)=v_{max}=\frac{P\beta}{2k}=\frac{6.75(0.5697)}{2(288)}$$

$$=5.61 \text{ mm}$$ ◄

$$M(0)=M_{max}=\frac{P}{4\beta}=\frac{6.75}{4(0.51697)}$$

$$=3.26421 \text{ kN}\cdot\text{m}$$

Thus,

$$\sigma_{max}=\frac{Mc}{I}=\frac{3264.21(0.0625)}{5.04\times10^{-6}}$$

$$=41.13 \text{ MPa}$$ ◄

9.12

From Eq. (9.3),

$$\beta=\left[\frac{18(10^3)}{4(0.375\times200\times10^9)(5.2\times10^{-7})}\right]^{1/4}$$

$$=0.824 \text{ m}^{-1}$$

Since

$$\beta L=0.824(0.74)=0.618<\pi/4$$

the beam can be considered rigid.

(a) Uniform deflection is

$$v=\frac{F}{3K}=\frac{540}{3(18)}=10 \text{ mm}$$ ◄

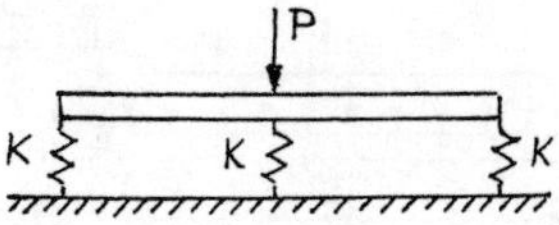

(CONT.)

9.12 CONT.

(b) We may replace the given beam by the beams shown in Figs.(a) and (b).

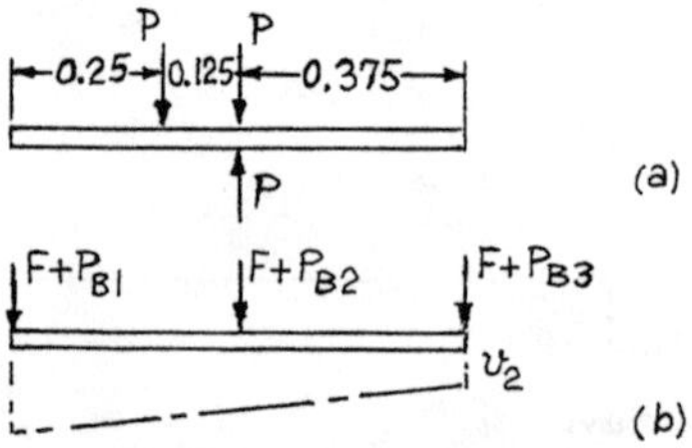

Note that F is the direct load on each spring and P_{Bi} is the bending loads with the values:

$$F = P/3 = 540/3 = 180 \text{ N}$$

and

$$P_{Bi} = \frac{M_B r_i}{\Sigma r_i^2} \qquad (i=1,2,3)$$

Here

$$P_{B1} = \frac{540(0.125)(0.375)}{2(15\times 0.125)^2} = 90 \text{ N}$$

$$P_{B2} = 0 \qquad P_{B3} = -P_{B1}$$

End deflections are thus,

$$v_1 = (180+90)/18 = 15 \text{ mm}$$
$$v_2 = (180-90)/18 = 5 \text{ mm}$$ ◄

9.13

Note: Slope may be found as in Prob.9.14, with p=0.

Using Eq. (9.3),

$$\beta = [14/4(8.4)]^{1/4} = 0.8034 \text{ m}^{-1}$$

$$\beta L = 0.8034(0.6) = 0.482 \text{ rad} = 27.62^\circ$$

Hence, from Eq. (9.13):

$$v_C = \frac{4500(0.8034)}{2(14\times 10^6)}\;\frac{2+\cos 27.62^\circ + \cosh 27.62^\circ}{\sin 27.62^\circ + \sinh 27.62^\circ}$$

$$= 186(10^{-6}) \text{ m} = 0.186 \text{ mm}$$ ◄

9.14

$\beta L < 1$

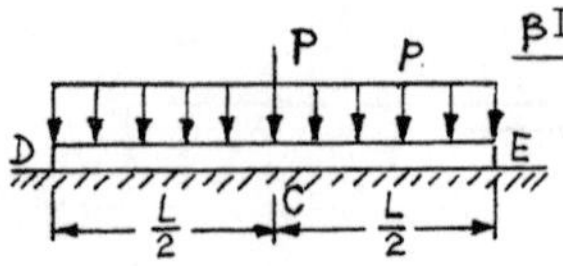

Referring to Sec. 9.6:

(CONT.)

9.14 CONT.

$$v_C = \frac{PL^3}{48EI} + \frac{5pL}{384EI} - \frac{5[(P+pL)/L]L^4}{384EI}$$

$$= \frac{(0.15)^3}{48(4.8\times 10^{-6})}\left[9000 + \frac{5(7500\times 0.15)}{8} - \frac{5(9000\times 0.15)}{8}\right]$$

$$= 2.81(10^{-8}) \text{ m}$$ ◄

Similarly,

$$\theta_D = \frac{PL^2}{16EI} + \frac{pL^3}{24EI} - \frac{[(9000+1125)/0.15](0.15)^3}{24EI}$$

Substituting the given data,

$$\theta_D = 5(10^{-7}) \text{ rad}$$ ◄

9.15

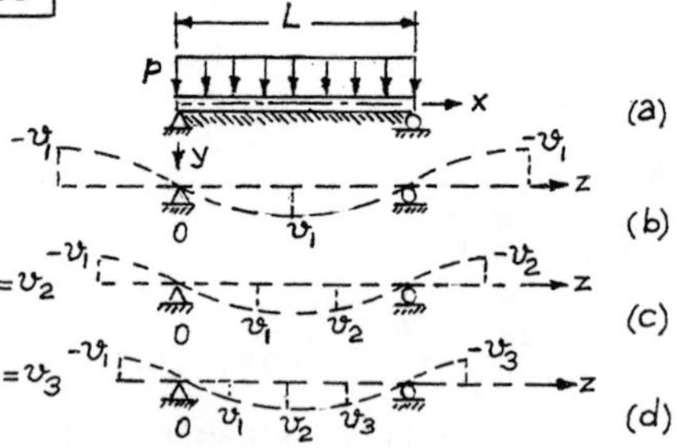

Boundary conditions are:

$$v(0)=v(L)=0, \quad v''(0)=v''(L)=0 \qquad \text{(a)}$$

These are transformed into the following central difference conditions by using Eqs. (7.6) and (7.9):

$$v_0=0, \quad -v_{-1}=v_1, \quad v_n=0, \quad v_{n-1}=-v_{n+1} \qquad \text{(b)}$$

For m=2 (Fig.b):

$$v_{n-2} - 4v_{n-1} + 6\left(\frac{m^4+1}{m^4}\right)v_n - 4v_{n+1} + v_{n+2} = \frac{1.6}{m^4}$$

or

$$-v_1 + 6\frac{2^4+1}{2^4}v_1 - v_1 = \frac{1.6}{2^4}; \qquad v_1 = 21.4 \text{ mm}$$ ◄

For m=3 (Fig.c):

$$-4v_1 + 6\frac{3^4+1}{3^4}v_2 - v_2 = \frac{1.6}{3^4}$$

$$-v_1 + 6\frac{3^4+1}{3^4}v_1 - 4v_2 = \frac{1.6}{3^4}$$

Solving, $v_1 = v_2 = 17.2$ mm ◄

(CONT.)

9.15 CONT.

For m=4 (Fig.d):

$$-v_1+6\frac{4^4+1}{4^4}v_1-4v_1+v_1=\frac{1.6}{4^4}$$

$$-4v_1+6\frac{4^4+1}{4^4}v_2-4v_1=\frac{1.6}{4^4}$$

Solving,

$$v_3=v_1=10.2\text{ mm}$$
$$v_2=13.9\text{ mm}$$ ◄

9.16

From Example 9.1:

$$v_p=\frac{p}{2k}(2-e^{-\beta a}\cos\beta a-e^{-\beta b}\cos\beta b)$$

where

$$k=48EI/aL_t^3$$

Thus,

$$v_p=\frac{paL_t^3}{2(48)EI}[2-e^{-\beta a}\cos\beta a-e^{-\beta b}\cos\beta b)$$

$$=\frac{paL_t^3}{96EI}[2-f_4(\beta a)-f_4(\beta b)]$$

We have

$$\beta=3.936/6=0.656\ \text{m}^{-1}$$

At midspan, we have

$$\beta a=\beta b \quad\text{and}\quad f_4(\beta a)\approx 0$$

Thus,

$$v_p=paL_t^3/48EI$$

Since,

$$v_m=\frac{paL_t^3}{48EI}=\frac{R_{cc}L_t^3}{48EI}$$

Solving,

$$R_{cc}=ap=0.3p$$ ◄

10.1

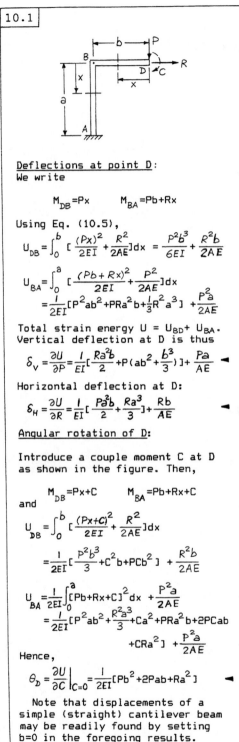

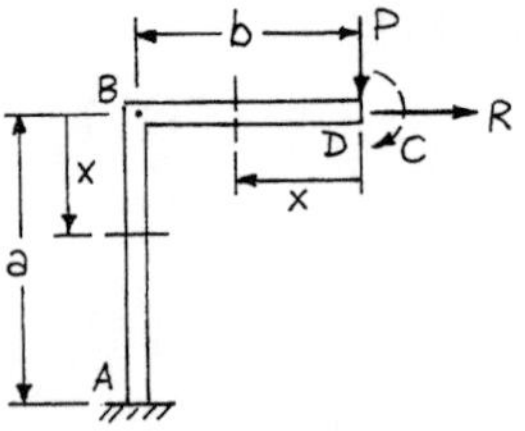

Deflections at point D:
We write

$$M_{DB}=Px \qquad M_{BA}=Pb+Rx$$

Using Eq. (10.5),

$$U_{DB}=\int_0^b \left[\frac{(Px)^2}{2EI}+\frac{R^2}{2AE}\right]dx = \frac{P^2b^3}{6EI}+\frac{R^2b}{2AE}$$

$$U_{BA}=\int_0^a \left[\frac{(Pb+Rx)^2}{2EI}+\frac{P^2}{2AE}\right]dx$$
$$=\frac{1}{2EI}[P^2ab^2+PRa^2b+\tfrac{1}{3}R^2a^3]+\frac{P^2a}{2AE}$$

Total strain energy $U = U_{BD}+ U_{BA}$. Vertical deflection at D is thus

$$\delta_V=\frac{\partial U}{\partial P}=\frac{1}{EI}\left[\frac{Ra^2b}{2}+P(ab^2+\frac{b^3}{3})\right]+\frac{Pa}{AE} \quad ◄$$

Horizontal deflection at D:

$$\delta_H=\frac{\partial U}{\partial R}=\frac{1}{EI}\left[\frac{Pa^2b}{2}+\frac{Ra^3}{3}\right]+\frac{Rb}{AE} \quad ◄$$

Angular rotation of D:

Introduce a couple moment C at D as shown in the figure. Then,

$$M_{DB}=Px+C \qquad M_{BA}=Pb+Rx+C$$

and

$$U_{DB}=\int_0^b \left[\frac{(Px+C)^2}{2EI}+\frac{R^2}{2AE}\right]dx$$
$$=\frac{1}{2EI}\left[\frac{P^2b^3}{3}+C^2b+PCb^2\right]+\frac{R^2b}{2AE}$$

$$U_{BA}=\frac{1}{2EI}\int_0^a[Pb+Rx+C]^2dx+\frac{P^2a}{2AE}$$
$$=\frac{1}{2EI}\left[P^2ab^2+\frac{R^2a^3}{3}+Ca^2+PRa^2b+2PCab+CRa^2\right]+\frac{P^2a}{2AE}$$

Hence,

$$\theta_D=\frac{\partial U}{\partial C}\Big|_{C=0}=\frac{1}{2EI}[Pb^2+2Pab+Ra^2] \quad ◄$$

Note that displacements of a simple (straight) cantilever beam may be readily found by setting b=0 in the foregoing results.

10.2

Equations of statics are applied to obtain the reactions:

$$\Sigma F_x=0: \qquad R_{Ax}=0$$
$$\Sigma M_A=0: \qquad R_C=\tfrac{1}{4}P_1+\tfrac{1}{2}P_2+\tfrac{3}{4}P_3$$
$$\Sigma M_C=0: \qquad R_{Ay}=\tfrac{3}{4}P_1+\tfrac{1}{2}P_2+\tfrac{1}{4}P_3$$

The axial force in each member is obtained by applying the method of sections, as required. The results are as follows:

$$N_{DC}=\frac{\sqrt{2}}{2}P_1+\frac{\sqrt{2}}{2}P_2+\frac{3\sqrt{2}}{4}P_3$$
$$N_{AE}=\frac{3\sqrt{2}}{4}P_1+\frac{\sqrt{2}}{2}P_2+\frac{\sqrt{2}}{4}P_3$$
$$N_{DE}=\frac{1}{2}P_1+P_2+\frac{1}{2}P_3$$
$$N_{BE}=-\frac{\sqrt{2}}{4}P_1+\frac{\sqrt{2}}{2}P_2+\frac{\sqrt{2}}{4}P_3$$
$$N_{BD}=\frac{\sqrt{2}}{4}P_1+\frac{\sqrt{2}}{2}P_2-\frac{\sqrt{2}}{4}P_3$$
$$N_{AB}=\frac{3}{4}P_1+\frac{1}{2}P_2+\frac{1}{4}P_3$$
$$N_{BC}=\frac{1}{4}P_1+\frac{1}{2}P_2+\frac{3}{4}P_3$$

Numerical results are determined for $P_1=P_2=P_3=45$ kN and L=3 m. These are tabulated below.

Bar	Axial force (kN)	$\partial N/\partial P_2$	Length (m)
DC	N_{DC} =99.5	$\sqrt{2}/2$	2.125
AE	N_{AE} =99.5	$\sqrt{2}/2$	2.125
DE	N_{DE} =90	1	3.000
BE	N_{BE} =31.8	$\sqrt{2}/2$	2.125
BD	N_{BD} =31.8	$\sqrt{2}/2$	2.125
AB	N_{AB} =67.5	1/2	3.000
BC	N_{BC} =67.5	1/2	3.000

Thus, the vertical deflection at B:

$$\delta_B=\frac{1}{AE}\sum_{i=1}^{7}N_i\frac{\partial N_i}{\partial P_2}L_i$$
$$=[2(67.5)(0.5)(3)+2(31.8)(0.707)(2.125)+90(1)(3)+2(99.5)(0.707)2.125](10^3)/AE$$

$$=686{,}773.675/AE \text{ m} \quad ◄$$

10.3

The moment is expressed by

$$M = P(2a-x); \quad \partial M/\partial P = 2a-x$$

Castigliano's theorem gives then

$$v_P = \int \frac{M}{EI}\frac{\partial M}{\partial P}\,dx$$

$$= \int_0^a \frac{P}{E}(2a-x)^2(c_1x+c_2)\,dx + \int_0^{2a}\frac{P}{EI_2}(2a-x)^2\,dx$$

or

$$v_P = \frac{11Pc_1a^4}{12E} + \frac{7Pc_2a^3}{3E} + \frac{Pa^3}{3EI_2}$$ ◄

10.4

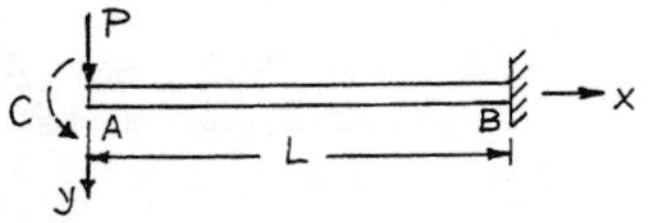

<u>Deflection at A</u>

$$M = Px$$

and

$$v_A = \frac{1}{EI}\int M\frac{\partial M}{\partial P}\,dx = \frac{1}{EI}\int_0^L Px(x)\,dx$$

$$v_A = \frac{PL^3}{3EI}$$ ◄

<u>Slope at A</u>

$$M = Px + C$$

and

$$\theta_A = \frac{1}{EI}\int M\frac{\partial M}{\partial C}\,dx = \frac{1}{EI}\int_0^L (Px+C)(1)\,dx$$

Setting C=0 and integrating

$$\theta_A = \frac{PL^2}{2EI}$$ ◄

10.5

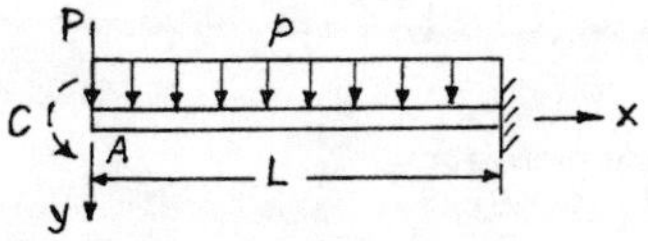

<u>Deflection at A</u>

$$M = Px + \tfrac{1}{2}px^2$$

and

$$\delta_A = \frac{1}{EI}\int M\frac{\partial M}{\partial P}\,dx = \frac{1}{EI}\int_0^L \left(Px+\tfrac{1}{2}px^2\right)(x)\,dx$$

$$= \frac{PL^3}{3EI} + \frac{pL^4}{8EI}$$ ◄

(CONT.)

10.5 CONT.

<u>Slope at A</u>

$$M = Px + \tfrac{1}{2}px^2 + C$$

and

$$\theta_A = \frac{1}{EI}\int M\frac{\partial M}{\partial C}\,dx = \frac{1}{EI}\int_0^L \left(Px+\tfrac{1}{2}px^2+C\right)dx$$

Setting C=0 and integrating,

$$\theta_A = \frac{PL^2}{2EI} + \frac{pL^3}{6EI}$$ ◄

10.6

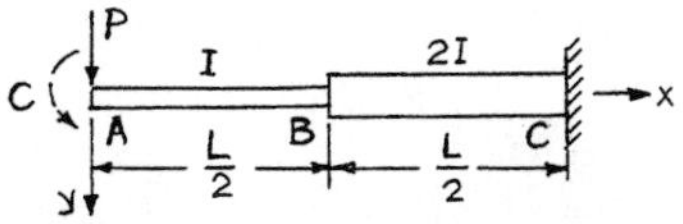

<u>Deflection at A</u>

$$M = Px$$

and

$$\delta_A = \int \frac{1}{EI} M\frac{\partial M}{\partial P}\,dx$$

$$= \frac{1}{EI}\int_0^{L/2} Px(x)\,dx + \frac{1}{2EI}\int_{L/2}^L Px(x)\,dx$$

Integrating,

$$\delta_A = \frac{PL^3}{24EI} + \frac{7PL^3}{48EI} = \frac{PL^3}{16EI}$$ ◄

<u>Slope at A</u>

$$M = Px + C$$

and

$$\theta_A = \int \frac{1}{EI} M\frac{\partial M}{\partial C}\,dx$$

$$= \frac{1}{EI}\int_0^{L/2}(Px+C)(1)\,dx + \frac{1}{2EI}\int_{L/2}^L (Px+C)\,dx$$

Setting C=0 and integrating,

$$\theta_A = \frac{PL^2}{8EI} + \frac{3PL^2}{16EI}$$

or

$$\theta_A = \frac{5PL^2}{16EI}$$ ◄

10.7

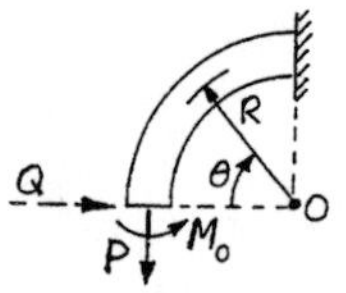

$ds = Rd\theta$

(a)

<u>Horizontal deflection</u>

$$M_\theta = PR(1-\cos\theta) + QR\sin\theta + M_o$$

and

$$\delta_H = \frac{1}{EI}\int M_\theta \frac{\partial M_\theta}{\partial Q} ds$$

$$= \frac{1}{EI}\int_0^{\pi/2} [PR^3(1-\cos\theta)\sin\theta + \overset{0}{\cancel{Q}}R^3\sin^2\theta + M_o R^2 \sin\theta]d\theta$$

$$= \frac{R^2}{2EI}\left(\frac{1}{2}PR + M_o\right)$$ ◄

<u>Vertical deflection</u> (Q=0)

$$\delta_V = \frac{1}{EI}\int M_\theta \frac{\partial M_\theta}{\partial P} ds$$

$$= \frac{1}{EI}\int_0^{\pi/2} [PR^3(1-\cos\theta)^2 + MR^2(1-\cos\theta)]d\theta$$

$$= \frac{R^2}{4EI}[PR(3\pi - 8) + 2M_o(\pi - 2)$$ ◄

(b) <u>Rotation of the free end</u> (Q=0)

$$\theta = \frac{1}{EI}\int M_\theta \frac{\partial M_\theta}{\partial M_o} ds$$

$$= \frac{1}{EI}\int_0^{\pi/2} [PR(1-\cos\theta) + M_o]d\theta$$

$$= \frac{R}{2EI}[PR(\pi - 2) + \pi M_o]$$ ◄

10.8

We now have

$M = -FR\sin\theta, \quad N = F\sin\theta, \quad V = -F\cos\theta$

Applying Eq. (10.6), with $\alpha = 6/5$ and $T=0$:

$$\delta_H = \frac{R^3}{EI}\int_0^{\pi} F\sin^2\theta\, d\theta + \frac{R}{AE}\int_0^{\pi} F\sin^2\theta\, d\theta + \frac{6R}{5AG}\int_0^{\pi} F\cos^2\theta\, d\theta$$

Integrating,

$$\delta_H = \frac{\pi F R^3}{2EI} + \frac{\pi F R}{2EA} + \frac{3\pi F R}{5AG}$$ ◄

(CONT.)

10.8 CONT.

Substituting the given data:

$$\delta_H = \frac{\pi(4000)(0.05)^3}{400(6.67)} + \frac{\pi(4000)(0.05)}{400(10^5)2} + \frac{3\pi(4000)(0.05)}{5(80\times 10^5)2}$$

$$= (0.59 + 0.008 + 0.02)10^{-3}$$
$$= 0.62 \text{ mm}$$ ◄

The error, if N and V are omitted, is 4.5%.

10.9

Introducing a rightward horizontal force Q at point D, we write

$M_1 = -Qx$	$0 \le x \le a$
$M_2 = -\frac{1}{2}Fx - Qa$	$0 \le x \le \frac{b}{2}$
$M_3 = -\frac{1}{2}Fx - Qa + F(x - \frac{b}{2})$	$\frac{b}{2} \le x \le b$
$M_4 = -\frac{1}{2}Fb + \frac{1}{2}Fb - Q(a-x)$	$0 \le x \le a$

Applying Castigliano's theorem, after setting Q=0, we have

$$\delta_{DH} = \frac{1}{EI}\int_0^{b/2} \frac{1}{2}Fax\,dx + \frac{1}{EI}\int_{b/2}^{b}\left(-\frac{1}{2}Fax + \frac{1}{2}Fab\right)dx$$

Integrating,

$$\delta_{DH} = \frac{Fab^2}{8EI}$$ ◄

10.10

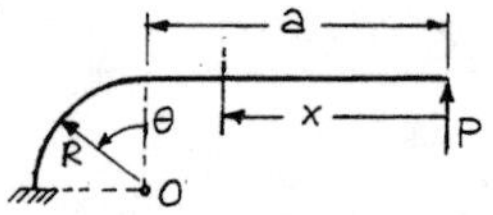

From symmetry,

$M_1 = -Px \qquad 0 \le x \le a$

$M_2 = -Pa - PR\sin\theta \qquad 0 \le \theta \le \pi/2$

Hence,

$$\frac{\delta}{2} = \frac{1}{EI}\int_0^a M_1 \frac{\partial M_1}{\partial P} dx + \frac{1}{EI}\int_0^{\pi/2} M_2 \frac{\partial M_2}{\partial P} R d\theta$$

or

$$\delta = \frac{2}{EI}\left[\int_0^a Px^2\, dx + P(a + R\sin\theta)^2 R d\theta\right.$$

$$= \frac{P}{6EI}(4a^3 + 6\pi R a^2 + 24R^2 a + 3\pi R^3)$$ ◄

10.11

Referring to Fig. P10.11, we write

$M=PR\sin\theta \qquad T=PR(1-\cos\theta)$

$m=R\sin\theta \qquad t=R(1-\cos\theta)$

Here t denotes torque caused by a unit load.

Applying Eq. (10.14), deflection at the free end (perpendicular to the plane of the ring):

$$\delta = \frac{1}{EI}\int_0^{\pi/2}(PR^2\sin^2\theta)Rd\theta + \frac{1}{JG}\int_0^{\pi/2}[PR^2(1-\cos\theta)^2 Rd\theta$$

$$= \frac{PR^3}{EI}\left|\frac{\theta}{2}-\frac{\sin 2\theta}{4}\right|_0^{\pi/2} + \frac{PR^3}{JG}\left|\theta-2\sin\theta+\frac{\theta}{2}+\frac{\sin 2\theta}{4}\right|_0^{\pi/2}$$

$$= \frac{PR^3}{4}[\frac{\pi}{EI}+\frac{3\pi-8}{JG}]$$

Letting $J=2I=\pi r^4/2$, we have

$$\delta = \frac{PR^3}{r^4}\left(\frac{1}{E}+\frac{0.226}{G}\right) \quad ◄$$

10.12

Referring to Fig. P10.12:

$M = -PR\sin\theta \qquad m=-R\sin\theta$

$T = PR(1-\cos\theta) \qquad t=R(1-\cos\theta)$

where t denotes torque caused by a unit load. We have

$\sin(2\pi-\theta)=-\sin\theta$

$\cos(2\pi-\theta)=\cos\theta$

Applying Eq. (10.14), deflection at the free end (perpendicular to the plane of the ring):

$$\delta = \frac{PR^3}{EI}\int_0^{2\pi}\sin^2\theta d\theta + \frac{PR^3}{JG}\int_0^{2\pi}(1-\cos\theta)^2 d\theta$$

Integrating,

$$\delta = PR^3\pi[\frac{1}{EI}+\frac{3}{JG}] \quad ◄$$

10.13

We introduce a couple moment C at point B. The expression for the moments are then,

$M_1 = -R_D x \qquad 0\le x\le \frac{L}{2}$

$M_2 = -R_D x+P(x-\frac{L}{2})+C \qquad \frac{L}{2}\le x\le L$

Applying Eq. (10.6):

$$\delta_B = \frac{1}{EI}\left\{\int_{L/2}^{L}[-R_D x + P(x-\frac{L}{2})+C] \times (x-\frac{L}{2})dx\right\}$$

Setting C=0 and integrating,

$$\delta_B = -\frac{PL^3}{48EI}\left[\frac{25/16}{1-(3EI/kL^3)}-2\right] \quad ◄$$

Similarly, applying Eq. (10.7):

$$\theta_B = \frac{1}{EI}\left\{\int_{L/2}^{L}[-R_D x + P(x-\frac{L}{2}) + C]dx\right\}$$

Setting C=0 and integrating,

$$\theta_B = -\frac{15PL^2}{128EI}\frac{1}{1-(3EI/kL^3)} \quad ◄$$

10.14

We write

$M_1 = Qx \qquad 0\le x\le a$

$M_2=Q(a+R\sin\theta)+PR(1-\cos\theta) \qquad 0\le\theta\le\pi$

Applying Eq. (10.6), with $\delta_Q=0$:

$$\frac{\partial U}{\partial Q}=0=\frac{1}{EI}\int_0^a Qx^2dx + \int_0^{\pi}[Q(a+R\sin\theta)+PR(1-\cos\theta)](a+R\sin\theta)Rd\theta$$

Integrating,

$$0 = \frac{QL^3}{3EI}+\frac{QR}{EI}(\pi a^2+4Ra+\frac{\pi}{2}R^2) + \frac{PR^2}{EI}(\pi a+2R)$$

This may be written in the following form:

$$Q = \frac{-PR^3(\pi a+2R)}{a^3[\frac{1}{3}+\frac{\pi R}{a}+4(\frac{R}{a})^2+\frac{\pi}{2}(\frac{R}{a})^3]} \quad ◄$$

10.15

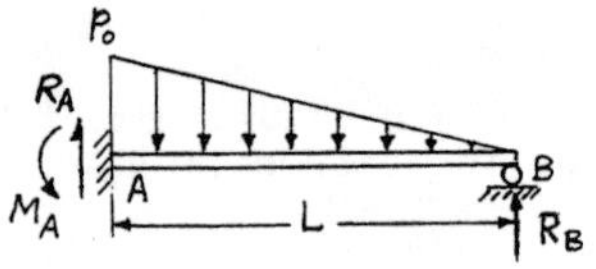

We have

$$M = -R_B x + \frac{p_o x}{L}\frac{x}{2}\frac{x}{3} = -R_B x + \frac{p_o x^3}{6}$$

Then,

$$\delta_B = 0 = \frac{1}{EI}\int M \frac{\partial M}{\partial R_B}dx = \frac{1}{EI}\int_0^L (R_B x - \frac{p_o x^3}{6L})x dx$$

This yields, after integrating,

$$R_B = p_o L/10 = R_r$$ ◄

Then, from equations of statics:

$$R_A = 4p_o L/10 = R_f$$

$$M_A = p_o L^2/15 = M_f$$ ◄

10.16

Moments are expressed by

$M_1 = -Rx \qquad 0 \le x \le \frac{L}{2}$

$M_2 = -Rx + M_o \qquad \frac{L}{2} \le x \le L$

Applying Eq. (10.6), with $\delta_R = 0$:

$$\frac{\partial U}{\partial R} = 0 = \frac{1}{EI}\int_0^{L/2} Rx^2 dx + \frac{1}{EI}\int_{L/2}^{L}(-Rx + M_o)(-x)dx$$

from which

$$R = 9M_o/8L$$ ◄

Using Eq. (10.7), slope at C:

$$\theta_C = \frac{\partial U}{\partial M} = 0 + \frac{1}{EI}\int_{L/2}^{L}(-Rx + M_o)dx$$

$$= \frac{5M_o L}{64EI}$$ ◄

10.17

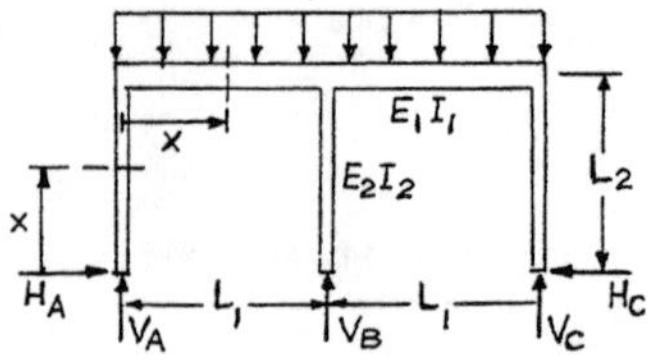

$M_1 = H_A x \qquad 0 \le x \le L_2$

$M_2 = -V_A x + H_A L_2 + \frac{1}{2}px^2 \qquad 0 \le x \le L_1$

(CONT.)

10.17 CONT.

Applying Eq. (10.6), with $\delta_{AV} = 0$ and $\delta_{AH} = 0$, respectively:

$$\frac{\partial U}{\partial V_A} = \frac{1}{E_1 I_1}\int_0^{L_1}(-V_A x + H_A L_2 + \frac{px^2}{2})(-x)dx = 0 \qquad (a)$$

$$\frac{\partial U}{\partial H_A} = \frac{1}{E_2 I_2}\int_0^{L_2} H_A x(x)dx + \frac{1}{E_1 I_1}\int_0^{L_1}(-V_A x + H_A L_2 + \tfrac{1}{2}px^2)L_2 dx = 0 \qquad (b)$$

Letting

$$\lambda = E_2 I_2 L_1 / E_1 I_1 L_2$$

Equations (a) and (b) become

$$8V_A - 12H_A L_2 = 3pL_1^2$$

$$3\lambda V_A L_1 - 2(3\lambda + 1)H_A L_2 = \lambda p L_1^2$$

or

$$V_A = \frac{3(\lambda+1)}{2(3\lambda+4)}pL_1 = R_{Av}$$

$$H_A = \frac{\lambda}{4(3\lambda+4)}\frac{pL_1^2}{L_2} = R_{AH}$$ ◄

The remaining reactions may then be found by using the equations of equilibrium.

10.18

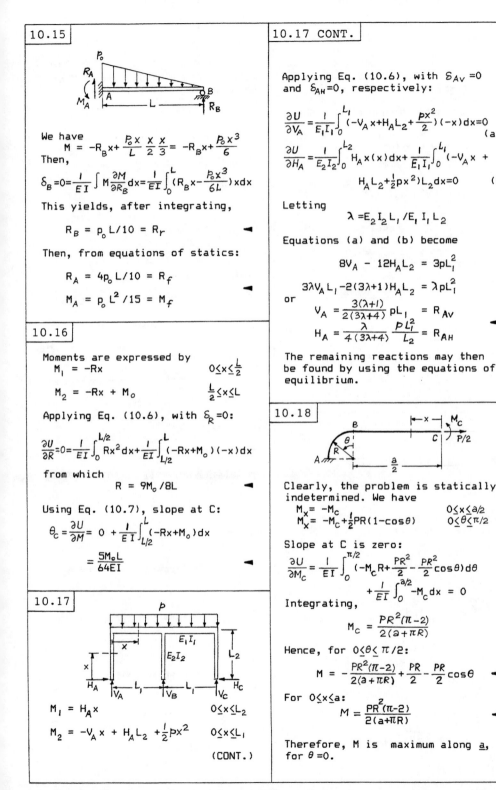

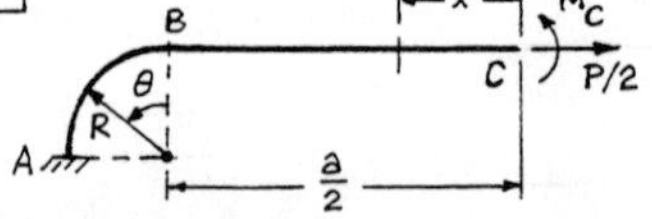

Clearly, the problem is statically indetermined. We have

$M_x = -M_C \qquad 0 \le x \le a/2$

$M_x = -M_C + \frac{1}{2}PR(1 - \cos\theta) \qquad 0 \le \theta \le \pi/2$

Slope at C is zero:

$$\frac{\partial U}{\partial M_C} = \frac{1}{EI}\int_0^{\pi/2}(-M_C R + \frac{PR^2}{2} - \frac{PR^2}{2}\cos\theta)d\theta + \frac{1}{EI}\int_0^{a/2} -M_C dx = 0$$

Integrating,

$$M_C = \frac{PR^2(\pi - 2)}{2(a + \pi R)}$$

Hence, for $0 \le \theta \le \pi/2$:

$$M = -\frac{PR^2(\pi-2)}{2(a+\pi R)} + \frac{PR}{2} - \frac{PR}{2}\cos\theta$$ ◄

For $0 \le x \le a$:

$$M = \frac{PR^2(\pi-2)}{2(a+\pi R)}$$ ◄

Therefore, M is maximum along a, for $\theta = 0$.

10.19

The complementary energy for the ith member of length L_i, from Eq. (2.39) is

$$U_{0i}^* = \int_0^{\sigma_i} \frac{\sigma_i^3}{K^3}\, d\sigma = \frac{\sigma_i^4}{4K^3} = \frac{1}{4K^3}\left(\frac{N_i}{A_i}\right)^4$$

The complementary energy of the truss is thus

$$U^* = \sum_{i=1}^{6} \frac{A_i L_i}{4K^3}\left(\frac{N_i}{A_i}\right)^4$$

Equation (10.16) is then

$$\delta_E = \sum_{i=1}^{6} \frac{L_i}{K^3}\left(\frac{N_i}{A_i}\right)^3 \frac{\partial N_i}{\partial P} \qquad \text{(P10.19)}$$

Applying the method of joints, as needed, we obtain

$N_1 = 3P/2$ $\quad N_2 = -5P/4$ $\quad N_3 = 0$

$N_4 = 5P/4$ $\quad N_5 = N_6 = -3P/4$

We have $L_1 = L_5 = L_6 = 3$ m, $L_3 = 4$ m, $L_2 = L_4 = 5$ m and $A_i = A$.

Equation (P10.19) becomes

$$\delta_E = \left(\frac{1}{AK}\right)^3 \left\{ L_1 N_1^3 \left(\frac{\partial N_1}{\partial P}\right) + L_2 N_2^3 \left(\frac{\partial N_2}{\partial P}\right) + \ldots + L_6 N_6^3 \left(\frac{\partial N_6}{\partial P}\right) \right\}$$

or

$$\delta_E = \left(\frac{P}{AK}\right)^3 \left\{ 3\left(\tfrac{3}{2}\right)^3\left(\tfrac{3}{2}\right) + 5\left(-\tfrac{5}{4}\right)^3\left(\tfrac{5}{4}\right) + 0 + \left(\tfrac{5}{4}\right)^3\left(\tfrac{5}{4}\right) + 3\left(\tfrac{3}{4}\right)^3\left(\tfrac{3}{4}\right) + 3\left(-\tfrac{3}{4}\right)^3\left(-\tfrac{3}{4}\right) \right\}$$

$$= \left(\frac{P}{AK}\right)^3 \frac{48 \times 81 + 5 \times 625 \times 2 + 3 \times 81 \times 2}{256}$$

or

$$\delta_E = 41.5(P/AK)^3 \quad ◄$$

10.20

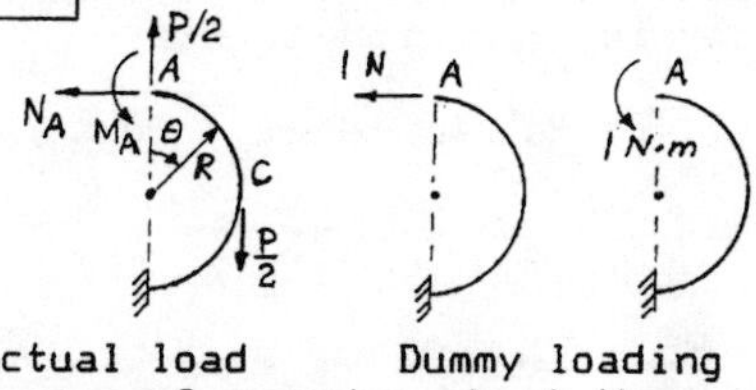

Actual load $\qquad$ Dummy loading

Because of symmetry about the vertical axis, only one half of the circle need be analyzed. Referring to the figure above, we have the following moments:

For $0 \le \theta \le \pi/2$:

$$M_1 = M_A + N_A R(1-\cos\theta) - \tfrac{1}{2}PR\sin\theta \qquad \text{(a)}$$

$$m_1 = R(1-\cos\theta), \quad m_1' = 1$$

(CONT.)

10.20 CONT.

For $\pi/2 \le \theta \le \pi$:

$$M_2 = M_A + N_A R(1-\cos\theta) - \tfrac{1}{2}PR \qquad \text{(b)}$$

$$m_2 = R(1-\cos\theta), \quad m_2' = 1$$

Horizontal deflection and slope are zero at A. Thus, from Eqs. (10.14) and (10.15):

$$\delta_{AH} = \frac{1}{EI}\int_0^{\pi} MR(1-\cos\theta)R\, d\theta = 0$$

or

$$\int_0^{\pi} M(1-\cos\theta)\, d\theta = 0 \qquad \text{(c)}$$

$$\theta_A = \frac{1}{EI}\int_0^{\pi} Mm'\, dx = \frac{1}{EI}\int_0^{\pi} M(1)dx = 0$$

or

$$\int_0^{\pi} M\, d\theta = 0 \qquad \text{(d)}$$

where M represents M_1 and M_2. Substituting Eq. (d) into Eq. (c), the latter reduces to

$$\int_0^{\pi} M\cos\theta\, d\theta = 0 \qquad \text{(e)}$$

Introducing Eqs. (a) and (b) into Eqs. (d) and (e), we have

$$M_A \int_0^{\pi} d\theta + N_A R \int_0^{\pi} (1-\cos\theta)\, d\theta - \tfrac{1}{2}PR\int_0^{\pi/2} \sin\theta\, d\theta - \tfrac{1}{2}PR\int_{\pi/2}^{\pi} d\theta = 0$$

and

$$M_A \int_0^{\pi} \cos\theta d\theta + N_A R \int_0^{\pi} (1-\cos\theta)\cos\theta d\theta - \tfrac{1}{2}PR\int_0^{\pi/2} \sin\theta\cos\theta d\theta - \tfrac{1}{2}PR\int_{\pi/2}^{\pi} \cos\theta d\theta = 0$$

Integrating and solving,

$$N_A = P/2\pi \qquad M_A = Pr/4 \quad ◄$$

10.21

$$L_{AD} = L_{DC} = 4.24 \text{ m}$$

$$U = \tfrac{1}{2} ALE\varepsilon^2 = \tfrac{1}{2} ALE(\delta_V \cos\alpha/L)^2$$

Vertical load of the joint, using Eq. (10.22),

$$\frac{\partial U}{\partial \delta_V} = P_V = \sum_{i=1}^{3} \frac{E_i A_i}{L_i}\, \delta_V \cos^2\alpha$$

or

$$P_V = EA\,\delta_V\left[\frac{\cos^2 45^\circ}{L_{AD}} + \frac{\cos^2 45^\circ}{L_{DC}} + \frac{1}{L_{BD}}\right]$$

$$= 0.5689 AE\,\delta_V$$

Substituting the numerical values,

$$P = 373.34 \text{ kN} \quad ◄$$

10.22

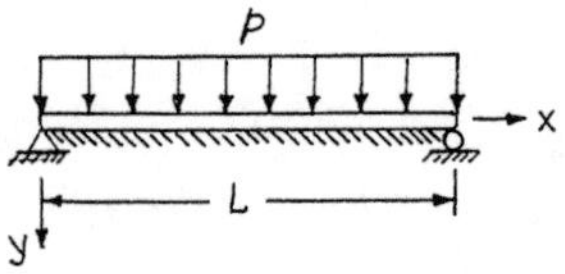

The deflection curve may be expressed by Eq. (10.23) and the bending strain energy is given by Eq. (10.25). The strain energy of deformation of the foundation is (Chap. 9):

$$U_2 = \frac{1}{2} k \int_0^L v^2 dx = \frac{1}{4} kL \sum_{n=1}^{\infty} a_n^2$$

Work done by the uniform load:

$$W = \frac{1}{2}\int_0^L p\, v\, dx = p\left(\frac{L}{\pi}\right) \sum_{n=1}^{\infty} \frac{1}{n}\, a_n \cos\frac{n\pi x}{L}\Big|_0^L$$

or

$$W = \frac{2pL}{\pi} \sum_{n=1,3,\ldots}^{\infty} \frac{a_n}{n}$$

Then, principle of virtual work yields

$$\frac{\pi^4 EI}{2L^3} n^4 a_n + \frac{kL}{2} a_n = \frac{2pL}{n\pi}$$

or

$$a_n = \frac{4pL^4}{n\pi(n^4\pi^4 EI + kL^4)}$$ ◄

Substituting this into Eq. (10.23) we obtain the required deflection curve.

10.23

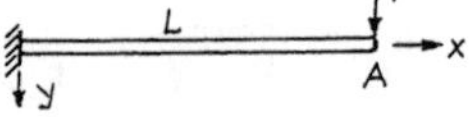

We obtain

$$v'' = 3\,\frac{a_1}{L^3}(L-x)$$

and

$$U = \frac{EI}{2}\int_0^L (v'')^2 dx = \frac{9EI a_1^2}{2L^6}\int_0^L (L-x)^2 dx$$

$$= \frac{3EI}{2L^3} a_1^2 \qquad \text{(a)}$$

Also $\qquad \delta W = P \cdot \delta v_A \qquad$ (b)

Thus, $\delta W = \delta U$ gives,

$$P \cdot \delta a_1 = \frac{3EI}{2L^3}(2a_1\, \delta a_1)$$

or

$$v_A = a_1 = PL^3/3EI$$ ◄

10.24

$$W = P \cdot v_A = P(a_1 L^2 + a_2 L^3)$$

$$= PL^2 (a_1 + a_2 L)$$

$$U = \frac{EI}{2}\int_0^L (v'')^2 dx$$

$$= 2EIL(a_1^2 + 3a_1 a_2 L + 3a_2^3 L^2)$$

Thus,

$$\frac{\partial \Pi}{\partial a_1} = 0: \quad PL^2 = 2EIL(2a_1 + 3a_2 L)$$

$$\frac{\partial \Pi}{\partial a_2} = 0: \quad PL^3 = 6EIL^2(a_1 + 2a_2 L)$$

Solving,

$$a_1 = PL/2EI \qquad a_2 = -P/6EI$$

Substituting back into Eq. (P10.24):

$$v = \frac{Px^2}{6EI}(3L - x)$$ ◄

10.25

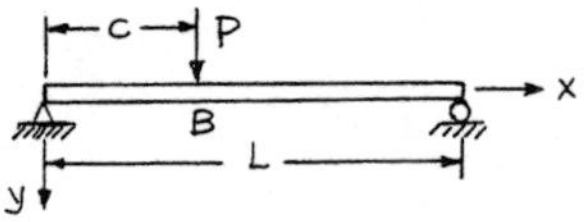

We have

$$\Pi = \frac{1}{2EI}\int_0^L (v'')^2 dx = P \cdot v_B \qquad \text{(a)}$$

Here

$$v = ax(L-x) = axL - ax^2$$

$$v' = aL - 2ax, \qquad v'' = -2a$$

Substituting these into Eq. (a), after integration, we obtain

$$\Pi = 2a^2 EIL - PacL + Pac^2$$

and

$$\frac{d\Pi}{da} = 4aEIL - PcL + Pc^2 = 0$$

or

$$a = \frac{Pc(L-c)}{4EIL}$$

The deflection of the beam at point B is therefore

$$v_B = \frac{Pc^2(L-c)^2}{4EIL}$$ ◄

10.26

The assumed deflection is of the general cubic form:

$$v = a_1 x^3 + a_2 x^2 + a_3 x + a_4 \qquad (a)$$

For the left hand portion of the beam:

$$v = 0 \quad \text{at} \quad x = 0 \qquad (1)$$
$$v' = 0 \quad \text{at} \quad x = 0 \qquad (2)$$
$$v' = 0 \quad \text{at} \quad x = L/2 \qquad (3)$$
$$v = \Delta \quad \text{at} \quad x = L/2 \qquad (4)$$

Here Δ is, deflection at midspan, to be determined.

Eqs. (1) and (2) give $a_3 = a_4 = 0$

Eq. (3) yields $a_2 = -3a_1 L/4$

Eq. (4) gives $a_1 = -16\Delta/L^3$

Introducing these into Eq. (a):

$$v = \frac{4\Delta x^2}{L^3}(3L - 4x) \qquad 0 \le x \le L/2 \qquad (b)$$

Strain energy is

$$U = 2\left(\frac{EI/2}{2}\right)\int_0^{L/4} (v'')^2 dx + 2\left(\frac{EI}{2}\right)\int_{L/4}^{L/2} (v'')^2 dx \qquad (c)$$

Substituting Eq. (b) into Eq. (c) and integrating, we obtain

$$U = 72EI\Delta^2/L^3$$

We have

$$\Pi = \frac{72EI\Delta^2}{L^3} - P\Delta$$

The midspan deflection, from $\partial\Pi/\partial\Delta = 0$, is thus

$$\Delta = PL^3/144EI = v_{max}$$

10.27

We obtain

$$v'' = \sum_{n=1}^{\infty} a_n \left(\frac{n\pi}{L}\right)^2 \cos\frac{n\pi x}{L}$$

$$U = \frac{EI}{2}\int_0^L (v'')^2 dx = \frac{EI}{2}\sum_{n=1}^{\infty} a_n^2 \left(\frac{n\pi}{L}\right)^4 \frac{L}{2}$$

$$W = P(v)_{x=\frac{L}{2}} = P\sum_n^{\infty} 2a_n \qquad (n=2,4,6,\ldots)$$

(CONT.)

10.27 CONT.

From the minimizing condition, $\partial\Pi/\partial a_n = 0$, we obtain

$$EIa_n\left(\frac{n\pi}{L}\right)^4 \frac{L}{2} - 2P = 0$$

or

$$a_n = \sum_n^{\infty} \frac{4PL^3}{n^4\pi^4 EI} \qquad (n=2,4,6,\ldots)$$

Therefore,

$$v_{max} = v\left(\frac{L}{2}\right) = \frac{8PL^3}{\pi^4 EI}\sum_n^{\infty}\frac{1}{n^4}$$

$(n=2,4,6,\ldots)$

11.1

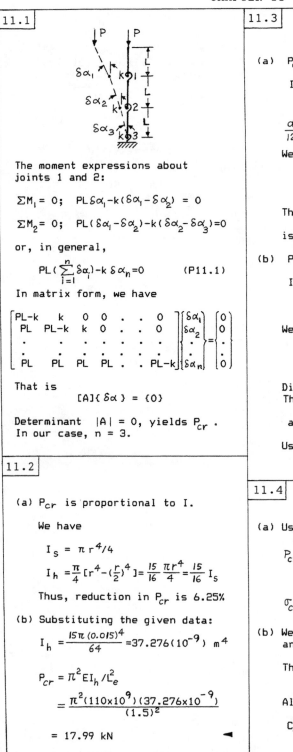

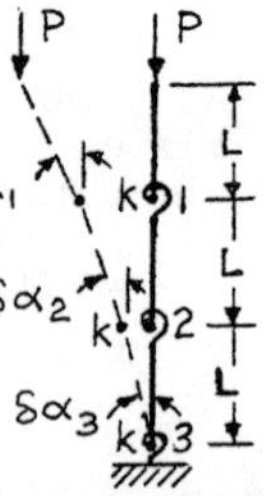

The moment expressions about joints 1 and 2:

$\Sigma M_1 = 0;\quad PL\delta\alpha_1 - k(\delta\alpha_1 - \delta\alpha_2) = 0$

$\Sigma M_2 = 0;\quad PL(\delta\alpha_1 - \delta\alpha_2) - k(\delta\alpha_2 - \delta\alpha_3) = 0$

or, in general,

$$PL\left(\sum_{i=1}^{n}\delta\alpha_i\right) - k\,\delta\alpha_n = 0 \qquad \text{(P11.1)}$$

In matrix form, we have

$$\begin{bmatrix} PL-k & k & 0 & 0 & . & . & 0 \\ PL & PL-k & k & 0 & . & . & 0 \\ . & . & . & . & . & . & . \\ . & . & . & . & . & . & . \\ PL & PL & PL & PL & . & . & PL-k \end{bmatrix} \begin{Bmatrix} \delta\alpha_1 \\ \delta\alpha_2 \\ . \\ . \\ \delta\alpha_n \end{Bmatrix} = \begin{Bmatrix} 0 \\ 0 \\ . \\ . \\ 0 \end{Bmatrix}$$

That is

$$[A]\{\delta\alpha\} = \{0\}$$

Determinant $|A| = 0$, yields P_{cr}. In our case, n = 3.

11.2

(a) P_{cr} is proportional to I.

We have

$I_s = \pi r^4/4$

$I_h = \frac{\pi}{4}[r^4 - (\frac{r}{2})^4] = \frac{15}{16}\frac{\pi r^4}{4} = \frac{15}{16} I_s$

Thus, reduction in P_{cr} is 6.25%

(b) Substituting the given data:

$I_h = \frac{15\pi(0.015)^4}{64} = 37.276(10^{-9})\ m^4$

$P_{cr} = \pi^2 E I_h / L_e^2$

$= \frac{\pi^2(110\times10^9)(37.276\times10^{-9})}{(1.5)^2}$

$= 17.99$ kN ◄

11.3

(a) $P_{cr} = 2(100) = 200$ kN

$I = \frac{P_{cr} L^2}{\pi^2 E} = \frac{200(10^3)(2)^2}{\pi^2(11\times10^9)}$

$= 7.369(10^6)\ mm^4$

$\frac{a^4}{12} = 7.369(10^6);\quad a = 96.97$ mm

We have $\sigma = \frac{P}{A} = \frac{100(10)^3}{(0.09697)^2}$

$= 10.63$ MPa < 15 MPa

Thus, a cross section of
97 x 97 mm ◄
is acceptable.

(b) $P_{cr} = 2(200) = 400$ kN

$I = \frac{400(10^3)(2)^2}{\pi^2(11\times10^9)} = 14.738(10^6)\ mm^4$

$= 14.738(10^6);\quad a = 115.32$ mm

We have $\sigma = \frac{P}{A} = \frac{200(10^3)}{(0.11532)^2}$

$= 15.04$ MPa > 15 MPa

Dimension is <u>not</u> acceptable. Therefore,

$a^2 = A = \frac{200(10^3)}{15(10^6)};\quad a = 115.5$ mm

Use a cross section of
116 x 116 mm ◄

11.4

(a) Using Eq. (11.5),

$P_{cr} = \frac{(200\times10^9)(0.075\times0.5^3/12)}{0.49(3.6)^2(2)}$

$= 127.2$ kN

$\sigma_{cr} = \frac{127.2\ (10^3)}{3.75(10^{-2})} = 33.91$ MPa ◄

(b) We have $I_{min} = 0.075(0.05)^3/12$ and

$r^2 = I_{min}/A = 2.08(10^{-4})$

Thus,

$\frac{L_e}{r} = \frac{0.7L}{r} = \frac{0.7(3.6)}{0.0144} = 175$

Also,

$C_c = \left[\frac{2\pi^2 E}{\sigma_{yp}}\right]^{1/2} = 121.673$

(CONT.)

11.4 CONT.

As 121.673<175, use Eq. (11.8):

$$\sigma_{all} = \frac{\pi^2(210\times10^9)}{1.92(175)^2} = 35.25 \text{ MPa}$$

Note that

$$\sigma = \frac{P}{A} = \frac{450(10^3)}{(0.05)(0.075)} = 0.12 \text{ MPa}$$

and yielding does not occur. But member fails as a column, since $P_{cr} < P$. ◄

11.5

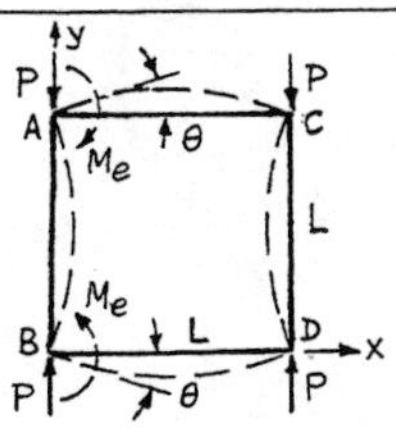

Symmetrical buckling shown in the figure, creates relative bending moments M_e which resist free rotation of the ends of the members AB and CD. Thus, for member AB:

$$EI\frac{d^2v}{dx^2} = -Pv + M_e$$

with the general solution

$$v = C_1\cos\lambda x + C_2\sin\lambda x + \frac{M_e}{P} \qquad (a)$$

where,

$$\lambda^2 = P/EI$$

Boundary conditions are

$$v(0) = 0 \qquad v'(\tfrac{L}{2}) = 0$$

$$v'(0) = \theta = M_eL/2EI$$

Introducing v from Eq. (a) into these:

$$C_1 + \frac{M_e}{P} = 0, \quad -C_1\lambda\sin\frac{\lambda L}{2} + C_2\lambda\cos\frac{\lambda L}{2} = 0$$

$$C_2\lambda = M_eL/2EI$$

These lead to the following trancental equation

$$\tan\frac{\lambda L}{2} + \frac{PL}{2\lambda EI} = 0$$

or

$$\tan\frac{\lambda L}{2} + \frac{\lambda L}{2} = 0$$

from which $\lambda L/2 = 2.029$.

(CONT.)

11.5 CONT.

Thus,

$$P_{cr} = \frac{16.47EI}{L^2} = \frac{\pi^2EI}{(0.774L)^2}$$

The effective length in the situation described is therefore equal to 0.774L. ◄

11.6

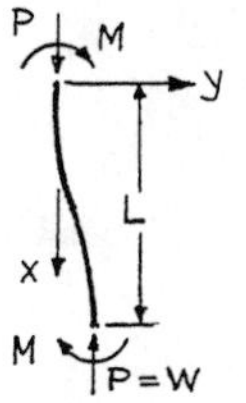

Let

$$\lambda_1 = P/EI$$
$$\lambda_2 = M/EI$$

The governing differential equation is

$$EI\,v'' = M - Pv$$

or

$$v'' + \lambda_1 v = \lambda_2^2$$

Solution is

$$v = A\cos\lambda_1 x + B\sin\lambda_2 x + (\frac{\lambda_2}{\lambda_1})^2$$

Boundary conditions give:

$$v'(0) = 0 = B$$

$$v(0) = A + (\frac{\lambda_2}{\lambda_1})^2 = 0; \quad A_1 = -(\frac{\lambda_2}{\lambda_1})^2$$

$$v'(L) = (\frac{\lambda_2}{\lambda_1})^2 \sin\lambda_1 L = 0$$

or

$$\sin\lambda_1 L = 0; \quad \lambda_1 L = 0, \pi, 2\pi, \ldots$$

Choose $\lambda_1 L = \pi$. Then,

$$\lambda_1 = \frac{\pi}{L} = (\frac{P}{EI})^{1/2}; \qquad P = \frac{\pi^2EI}{L^2}$$

Thus,

$$L_e = L$$ ◄

11.7

For the vertical bar, from Eq. (11.5):

$$P_{cr} = \pi^2EI/4L$$

For the midspan deflection of the beam is thus

$$\delta = \frac{5pL^4}{384EI} - \frac{P_{cr}L^3}{48EI}$$

or

$$\delta = \frac{5pL^4}{384EI} - \frac{\pi^2L}{192} = \frac{L}{192}(\frac{5pL^3}{2EI} - \pi^2)$$ ◄

11.8

(a) We have $\alpha(\Delta T)L = \delta/2$
or

$$\Delta T = \delta/2\alpha L \quad ◄$$

(b)

$$\alpha(\Delta T)L - \frac{PL}{AE} = \frac{\delta}{2}$$

where

$$P = \pi^2 EI/4L^2$$

Thus,

$$\Delta T = \frac{\delta}{2\alpha L} + \frac{\pi^2 EI}{4L^2}\,\frac{L}{AE}\,\frac{1}{\alpha L}$$

$$= \frac{\delta}{2\alpha L} + \frac{\pi^2 I}{4L^2 A\alpha} \quad ◄$$

11.9

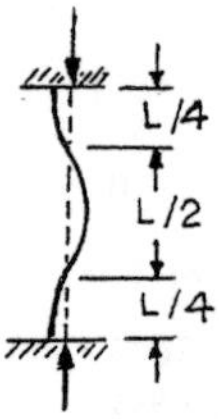

For a fixed ended column L_e may be determined as follows. Inflection points and midpoint divide the bar into 4 equal portions. Each portion is the same as the fundamental case (Fig. 11.2a). Thus, $L_e = L/4$.

For the dimension given, we obtain

$$\frac{L}{r} = \frac{1.932}{0.0294} = 65.99$$

Equation (11.6) gives then

$$\sigma_{cr} = \frac{4\pi^2 E}{(L/r)^2} = 0.0091E$$

Substituting $E = 174(10^9)$,

$\sigma_{cr} = 0.0091(175\times10^9) = 1592.5$ MPa

Since, from Fig. P11.9, E is valid only upto 175 MPa, the 1592.5 MPa cannot be critical stress.

Similarly, for inelastic range, from Eq. (11.7):

$\sigma_{cr} = 0.0091E_{t1}$
$= 0.0091(46.7\times10^9) = 425$ MPa

Applying the same reasoning as before, this value is also not a critical stress.

For E_{t2}:

$\sigma_{cr} = 0.0091(28\times10^9) = 254.8$ MPa

We now observe from the sketch that 254.8 falls in the stress range for which E_{t2} is valid. Thus, the buckling load:

$P_f = \sigma_{cr}A = 254.8(0.0323)$
$= 823$ kN

11.10

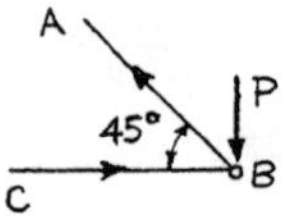

For bar AB:

$$1.41P = 4200(5.4\times10^{-5}) = 2268$$

or

$$P = 1604 \text{ N}$$

For bar BC:

$$P = P_{cr} = \pi^2 EI/L^2$$

$$= \frac{\pi^2(210\times10^9)(3.91\times10^{-9})}{6.25}$$

$$= 1297 \text{ N}$$

And

$$\sigma_{cr} = \frac{1297}{5.4(10^{-5})} = 24 \text{ MPa}$$

Conclusion: bar BC fails as a column, $P_{cr} = 1297$ N. ◄

11.11

Referring to Fig. P11.11, we write

$$I_y = 2Ar_c^2 + 2Ad^2$$
$$= 2(0.0225)^2 + 2(1.719\times10^{-3}) \times (0.0125+0.2325)^2$$
$$= 6.13(10^{-6}) \text{ m}^4$$

and

$$r_y = (I_y/2A)^{1/2} = (r_c^2+d^2)^{1/2}$$
$$= (0.0225^2 + 0.03575^2)^{1/2}$$
$$= 42.24 \text{ mm}$$

We see that r_y for two channels is the same as for one channel. But $r_z = r_c = 0.0225$ m. Thus, $r_z < r_y$. Columns tend to buckle with respect to z axis. The slenderness ratios:

$$\frac{L_e}{r_z} = \frac{2.10}{0.0225} \approx 94, \quad \frac{L_e}{r_z} = \frac{4.20}{0.0225} \approx 187$$

From Sec. 11.6:

$$C_c^2 = 2\pi^2 E/\sigma_{yp} = 2\pi^2(210\times10^9)/203(10^6)$$
$$= 20{,}420; \quad C_c = 143$$

(a) In this case $0 < (L_e/r) < C_c$. Then, substituting the given data, the first of Eqs. (11.8) yields

$$\sigma_{all} = 97.685 \text{ MPa} \quad ◄$$

(b) Now we have: $C_c \le (L_e/r) \le 200$ and the second of Eqs. (11.8) gives

$$\sigma_{all} = 30.87 \text{ MPa} \quad ◄$$

11.12

From Fig. P11.11, we observe that

$$I_y > I_z$$

and the column buckles with respect to the z axis. Since,

$$L/r = 4.2/0.0225 = 186.7$$

We have

$$\sigma_{cr} = \pi^2 E/(L_e/r)^2$$

or

$$= \frac{\pi^2(210\times10^9)}{(186.7)^2} = 59.46 \text{ MPa} \quad ◄$$

11.13

From Eq. (11.6),

$$\left(\frac{L}{r}\right)_{limit} = \left[\frac{\pi^2(140\times10^9)}{280\times10^6}\right]^{1/2} = 70.2$$

We have

$$I = (bh^3/12) = Ar^2$$

or

$$I = (0.05)(0.025)^3/12 = 1.25\times10^{-3} r^2$$

Solving,

$$r = 7.216 \text{ mm}$$

Hence,

$$\left(\frac{L}{r}\right)_{actual} = 1.2/7.216(10^{-3}) = 166.3$$

Thus, elastic buckling occurs, since $(L/r)_{limit} < (L/r)_{actual}$.

We have

$$\Delta = \alpha(\Delta T)L$$

and

$$\varepsilon = \frac{\Delta}{L} = \alpha(\Delta T)$$

The condition that

$$\varepsilon = \alpha(\Delta T) - \frac{\pi^2}{(L/r)^2} = 0$$

results in

$$\Delta T = \frac{\pi^2}{\alpha(L/r)^2} = \frac{\pi^2}{10(10^{-6})(166.3)^2}$$

$$= 35.7^\circ\text{C} \quad ◄$$

11.14

$$\sigma_{all} = \frac{P}{A} = \frac{125(10^3)}{3.06(10^{-3})} = 40.85 \text{ MPa}$$

Also

$$C_c = \left[\frac{2\pi^2(200\times10^9)}{250(10^6)}\right]^{1/2} = 125.7$$

Assuming $(L/r) \geq C_c$:

(CONT.)

11.14 CONT.

$$\sigma_{all} = \frac{\pi^2(200\times10^9)}{1.92(L/r)^2} = \frac{1028.08(10^9)}{(L/r)^2}$$

or

$$40.85(10^6) = \frac{1028.08(10^9)}{(L/r)^2}$$

Solving,

$$L/r = 158.6 \qquad \text{O.K.}$$

Choosing the smallest radii of gyration:

$$\frac{L}{r_y} = \frac{L}{0.0246} = 158.6$$

from which

$$L = 3.9 \text{ m} \quad ◄$$

11.15

We have

$$p = 77(10^3)(0.05\times0.05) = 192.5 \text{ N/m}$$

$$I = (0.05)^4/12 = 5.2(10^{-7}) \text{ m}^4$$

(a)

$$\sigma_{max} = \frac{Mc}{I} = \frac{192.5(9)^2}{8}\,\frac{0.025}{5.2(10^{-7})}$$

$$= 93.705 \text{ MPa} \quad ◄$$

$$v_{max} = \frac{5pL^4}{384EI}$$

$$= \frac{5(192.5)(9)^4}{384(210\times10^9)(5.2\times10^{-7})}$$

$$= 0.1506 \text{ m} = 150.6 \text{ mm} \quad ◄$$

(b) Taking

$$a_0 = 5pL^4/384EI = 150.6 \text{ mm}$$

We obtain, using Eq. (11.10):

$$v_{max} = \frac{0.1506}{1 - \dfrac{4500(9)^2}{(210\times10^9)(5.2\times10^{-7})}}$$

$$= 227.5 \text{ mm} \quad ◄$$

Applying Eq. (11.11):

$$\sigma_{max} = \frac{4500}{0.025}\left[1 + 0.2275\,\frac{0.025}{5.2(10^{-7})/0.025}\right]$$

$$= 51.019 \text{ MPa} \quad ◄$$

11.16

We have e = 0.05 m, c = 0.1016 m and

$$r_z^2 = I_z/A = \frac{45.66}{5880} = 7.765\ (10^{-3})\ \text{mm}^2$$

Then,

$$P_{cr} = \frac{\pi^2 EI_y}{L^2} = \frac{\pi^2(210\times10^3)15.4}{(4.5)^2} = 1575\ \text{kN}$$

Also,

$$P_{cr} = \frac{\pi^2(210\times10^3)45.66}{(4.5)^2} = 4669\ \text{kN}$$

Thus, Eq. (11.14) becomes

$$210(10^6) = \frac{P}{588(10^{-6})}\left[1+\frac{0.05(0.1016)}{7.765\,(10^{-3})} \times \sec\left(\frac{\pi}{2}\sqrt{\frac{P}{4669\times10^3}}\right)\right]$$

from which, by trial and error,

$$P \approx 700\ \text{kN} = P_{max}$$ ◄

11.17

Governing equation is

$$EI\,v_1'' = -(v_0+v_1)$$

where,

$$v_0 = a_1\sin\frac{\pi x}{L} + 5a_1\sin\frac{2\pi x}{L}$$

This may be written

$$v_1''+\lambda^2 v_1 = -\lambda^2 a_1\sin\frac{\pi x}{L} - 5\lambda^2 a_1\sin\frac{2\pi x}{L} \qquad (a)$$

We have

$$v_1 = c_1\cos\lambda x + c_2\sin\lambda x + v_p \qquad (b)$$

Particular solution is

$$v_p = A\sin\frac{\pi x}{L} + B\cos\frac{\pi x}{L} + D\sin\frac{2\pi x}{L} + E\cos\frac{2\pi x}{L} \qquad (c)$$

Substituting Eq. (c) into Eq. (a):

$$A = \lambda^2 a_1/(\frac{\pi^2}{L^2}-\lambda^2) \qquad B = 0$$

$$D = 5\lambda^2 a_1/(\frac{4\pi^2}{L^2}-\lambda^2) \qquad E = 0$$

Thus,

$$v_p = \frac{\lambda^2 a_1}{(\pi/L)^2-\lambda^2}\sin\frac{\pi x}{L} + \frac{5\lambda^2 a_1}{4(\pi/L)^2-\lambda^2}\sin\frac{2\pi x}{L}$$

Boundary conditions

$$v_1(0) = 0 \qquad v_1(L) = 0$$

yield $c_1 = c_2 = 0$.

(CONT.)

11.17 CONT.

General solution is

$$v = v_1 + v_p = v_0 + v_p$$

Letting

$$b = \frac{\pi^2}{\lambda^2 L^2} = \frac{PL^2}{\pi^2 EI}$$

we obtain

$$v = \frac{a_1}{1-b}\sin\frac{\pi x}{L} + \frac{20a_1}{4-b}\sin\frac{2\pi x}{L}$$

Solution of the b is found from

$$v(\tfrac{3L}{4}) = 0 = \frac{a_1}{1-b}\sin\frac{\pi(\frac{3L}{4})}{L} + 20\frac{a_1}{1-b}\sin\frac{2\pi(\frac{3L}{4})}{L}$$

or

$$b = 0.89$$

Thus,

$$b = \frac{PL^2}{\pi^2 EI}, \qquad P = \frac{b\pi^2 EI}{L^2}$$

and

$$P = 0.89\frac{\pi^2 EI}{L^2}$$ ◄

11.18

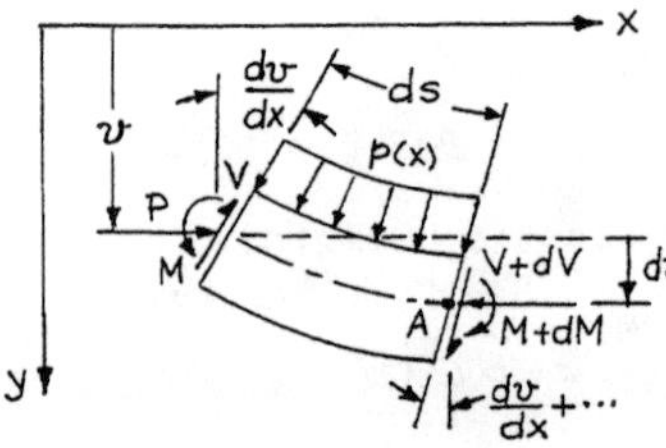

An element isolated from the beam is shown in a deformed state in the figure above. Assume $\sin\theta \approx \theta$, $\cos\theta \approx 1$, and $ds \approx dx$. Then

$$\Sigma F_y = 0: \quad -V + pdx + (V+dV) = 0$$

or

$$p = -dV/dx \qquad (a)$$

$$\Sigma M_A = 0: \quad M - Pdv - Vdx + pdx\frac{dx}{2} - (M+dM) = 0$$

Neglecting terms of second order, this becomes

$$V = -\frac{dM}{dx} - p\frac{dv}{dx} \qquad (b)$$

If the shear and axial deformations are neglected, the moment at any point is

$$M = EI\frac{d^2v}{dx^2} \qquad (c)$$

(CONT.)

11.18 CONT.

Substitution of Eqs. (c) and (b) into Eq. (a) gives

$$\frac{d^2}{dx^2}\left(EI\frac{d^2v}{dx^2}\right) + P\frac{d^2v}{dx^2} = p \qquad (d)$$

For EI = constant,

$$\frac{d^2v}{dx^2} + \frac{P}{EI}\frac{d^2v}{dx^2} = \frac{p}{EI} \qquad (c)$$

Homogeneous solution of this equation is

$$v = c_1 \sin\sqrt{\frac{P}{EI}}\,x + c_2\cos\sqrt{\frac{P}{EI}}\,x + c_3 x + c_4$$

where, c_1 through c_4 will require for evaluation, four boundary conditions.

11.19

Given $v = a_0[1-(4x^2/L^2)]$ and hence,

$$v' = -8xa_0/L^2 \qquad v'' = -8a_0/L^2$$

Potential energy function is

$$\Pi = 2\int_0^{L/2}\frac{1}{2}EI(v'')^2dx - 2\int_0^{L/2}\frac{1}{2}P(v')^2dx$$

$$= \frac{32EIa_0^2}{L^3} - \frac{32}{12}\frac{Pa_0^2}{L}$$

Thus,

$$\frac{\delta\Pi}{\delta a_0} = \frac{64EIa_0}{L^3} - \frac{64}{12}\frac{Pa_0}{L} = 0$$

or

$$P_{cr} = 12EI/L^2 \quad ◄$$

11.20

Assume $v = v_0\sin(\pi x/L)$. Then,

$$U = 2\int_0^{L/2} EI_1\left(1+\frac{x}{L/2}\right)(v'')^2dx$$

$$= 2EI_1 v_0^2\frac{\pi^4}{L^4}\left[\int_0^{L/2}\sin^2\frac{\pi x}{L}dx + \frac{2}{L}\int_0^{L/2}x\sin^2\frac{\pi x}{L}dx\right]$$

$$= 2EI_1 v_0^2\frac{\pi^4}{L^4}\left[\frac{L}{4} + \frac{2L}{\pi^2}\left(\frac{\pi^2}{16}+\frac{1}{4}\right)\right] \qquad (a)$$

We also have

(CONT.)

11.20 CONT.

$$\int_0^L (v')^2dx = v_0^2\frac{\pi^2}{L^2}\int_0^L\cos^2\frac{\pi x}{L}dx = v_0^2\pi^2/2L \qquad (b)$$

Thus,

$$\int_0^L EI(v'')^2dx = P\int_0^L (v')^2dx$$

yields

$$P_{cr} = \frac{\pi^2EI_1}{L^2}\left(\frac{3}{2}+\frac{2}{\pi^2}\right) = 1.7\frac{\pi^2EI_1}{L^2} \quad ◄$$

11.21

We have

$$v' = 2v_1\frac{x}{L^2} \qquad v'' = 2\frac{v_1}{L^2}$$

Potential energy function is

$$\Pi = \frac{EI_1}{2}\int_0^L\left(1-\frac{x}{2L}\right)\frac{4v_1^2}{L^4}dx - \frac{P}{2}\int_0^L\frac{4v_1^2}{L^4}x^2dx$$

$$= \frac{3}{2}\frac{EI_1v_1^2}{L^3} - \frac{2}{3}\frac{v_1^2P}{L}$$

Hence,

$$\frac{\delta\Pi}{\delta v_1} = \frac{3EI_1v_1}{L^3} - \frac{4v_1P}{3L} = 0$$

gives

$$P_{cr} = 9EI_1/4L^2 \quad ◄$$

11.22

Assume $v = \sum_1^\infty a_n\sin(n\pi x/L)$. Then,

$$U = \frac{\pi^4EI}{4L^3}\sum n^4a_n^4$$

and

$$W = \frac{1}{2}\int_0^L P(v')^2dx + \int_0^L pv\,dx$$

$$= \frac{\pi^2P}{4L}\sum n^2a^2 + \frac{2pL}{\pi}\sum a_n\frac{1}{n}$$

Hence, from $\delta U = \delta W$, by letting

$$b = \frac{P}{P_{cr}} = \frac{PL^2}{\pi^2EI}$$

We have

$$a_n = \frac{4pL^4}{\pi^5EI}\sum_{1,3,\ldots}^\infty\frac{1}{n^3(n^2-b)}$$

Thus,

$$v = \frac{4pL^4}{\pi^5EI}\sum_{1,3,\ldots}^\infty\frac{1}{n^3(n^2-b)}\sin\frac{n\pi x}{L} \quad ◄$$

11.23

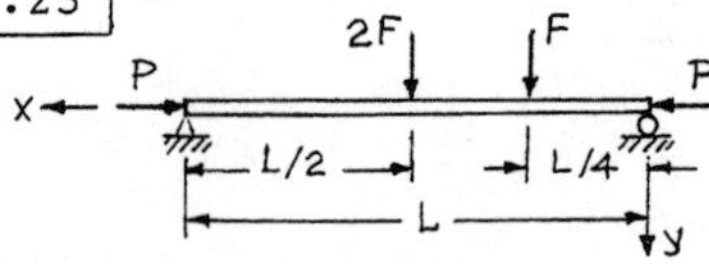

Let $c_1 = L/4$ $\quad c_2 = L/2$.
Deflection curve is expressed by

$$v = \sum_1^\infty a_n \sin\frac{n\pi x}{L} \qquad (a)$$

and

$$U = \frac{EI}{2}\int_0^L (v'')^2 dx$$

$$= F\sum a_n \sin\frac{n\pi c_1}{L} + 2F\sum a_n \sin\frac{n\pi c_2}{L} + \frac{P}{2}\int_0^L (v')^2 dx$$

Thus,

$$\partial(U-W)/\partial a_n = 0$$

gives

$$\frac{\pi^4 EI}{2L^3} n^4 a_n \delta a_n = F\delta a_n \sin\frac{n\pi c_1}{L} + 2F\delta a_n \sin\frac{n\pi c_2}{L} + \frac{P\pi^2}{2L} a_n \delta a_n n^2$$

or

$$a_n = \frac{F\sin(n\pi/4) + 2F\sin(n\pi/2)}{\frac{\pi^4 EI n^2}{2L^3}\left(n^2 - \frac{P}{P_{cr}}\right)} \quad ◄$$

Solution is found by substituting this into Eq. (a).

11.24

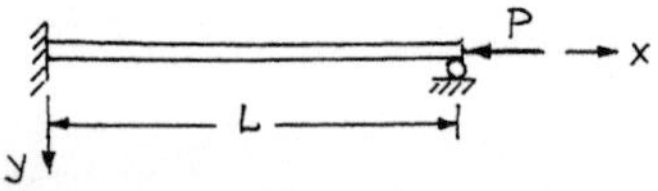

Boundary conditions are

$$v(0) = v(L) = 0$$

We have

$$v = aLx^2 + bLx^3 - ax^3 - bx^4$$

Thus,

$$v' = 2aLx + 3bLx^2 - 3ax^2 - 4bx^3$$

$$v'' = 2aL + 6BLx - 6ax - 12bx^2$$

Hence,

$$\Pi = \frac{EI}{2}\int_0^L (v'')^2 dx - \frac{P}{2}\int_0^L (v')^2 dx$$

$$= \frac{EI}{2}[4a^2L^3 + 8abL^4 + \frac{24}{5}b^2L^5] - \frac{P}{2}[\frac{2}{15}a^2L^5 + \frac{1}{5}abL^6 + \frac{3}{35}b^2L^7]$$

It is required that

$$\frac{\partial \Pi}{\partial a} = \frac{EI}{2}(8aL^3 + 8abL^4) - \frac{P}{2}(\frac{4}{15}aL^5 + \frac{bL^6}{5}) = 0$$

$$\frac{\partial \Pi}{\partial b} = \frac{EI}{2}(8aL^4 + \frac{48}{5}bL^5) - \frac{P}{2}(\frac{aL^6}{5} + \frac{bL^7}{35}) = 0$$

(CONT.)

11.24 CONT.

Letting $\lambda = PL^2/EI$, these become

$$(4 - \frac{2\lambda}{35})a + (4 - \frac{\lambda}{10})bL = 0$$

$$(4 - \frac{\lambda}{10})a + (\frac{24}{5} - \frac{3\lambda}{35})bL = 0$$

Since $a \neq 0$ and $b \neq 0$:

$$-(4 - \frac{\lambda}{10})^2 + (4 - \frac{2\lambda}{15})(\frac{24}{5} - \frac{3\lambda}{35}) = 0$$

or

$$\lambda^2 - 128\lambda + 2240 = 0$$

Solving,

$$\lambda_1 = 20.9 \qquad \lambda_2 = 107.1$$

Hence,

$$P_{cr} = \frac{20.9EI}{L} = 2.12\frac{\pi^2 EI}{L^2} \quad ◄$$

11.25

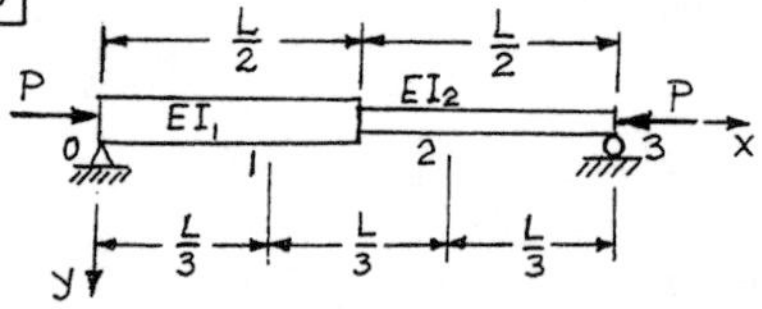

From Eq. (11.24):

$$v_{m+1} + (\lambda_1 h^2 - 2)v_m + v_{m-1} = 0 \qquad (a)$$

Here

$$\lambda_1^2 = P/EI_1 \qquad \lambda_2^2 = P/EI_2$$

Applying Eq. (a) at 1 and 2:

$$\begin{bmatrix} \lambda_1 h^2 - 2 & 1 \\ 1 & \lambda_2 h^2 - 2 \end{bmatrix}\begin{Bmatrix} v_1 \\ v_2 \end{Bmatrix} = \begin{Bmatrix} 0 \\ 0 \end{Bmatrix}$$

Thus,

$$(\lambda_1 h^2 - 2)(\lambda_2 h^2 - 2) - 1 = 0$$

or

$$\lambda_1\lambda_2 h^4 - 2\lambda_1 h^2 - 2\lambda_2 h^2 + 3 = 0$$

This is written as

$$P^2\left[\frac{L^4}{81E^2 I_1 I_2}\right] - P\left[\frac{2L^2}{9EI_1} + \frac{2L^2}{9EI_2}\right] + 3 = 0$$

Solution is

$$(P_{1,2})_{cr} = \frac{9E}{L^2}(I_1 + I_2) \pm \frac{9E}{L^2}[I_1^2 - I_1 I_2 + I_2^2]^{1/2}$$

When $I_1 = I_2$:

$$P_{cr} = \frac{9EI}{L^2} \quad ◄$$

11.26

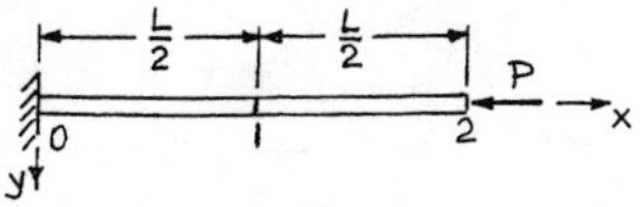

Boundary conditions yield

$$v'(0) = 0; \qquad v_0 = 0$$
$$v(0) = 0; \qquad v_1 = v_{-1}$$
$$v''(L) = 0; \qquad v_3 - 2v_2 + 2v_1 = 0 \qquad (1)$$
$$v'''(L) = 0; \qquad v_4 - 2v_3 + 2v_1 = 0 \qquad (2)$$

Applying Eq. (1) of Example 11.9 at points 1,2,0, respectively:

$$(7-2\lambda^2h^2)v_1 + (\lambda^2h^2-4)v_2 + v_3 = 0 \qquad (3)$$

$$(\lambda^2h^2-4)v_1 + (6-2\lambda^2h^2)v_2 + (\lambda^2h^2-4)v_3 + v_4 = 0 \qquad (4)$$

$$(\lambda^2h^2-4)v_1 + v_2 = 0 \qquad (5)$$

From Eq. (5):

$$v_2 = -v_1(\lambda^2h^2-4)$$

Equation (1) becomes

$$v_3 = 2v_2 - v_1 = -2v_1(\lambda^2h^2-4) - v_1$$

Substituting this into Eq. (3),

$$(7-2\lambda^2h^2)v_1 - (\lambda^2h^2-4)(\lambda^2h^2-4)v_1 - 2v_1(\lambda^2h^2-4) - v_1 = 0$$

from which

$$\lambda^4h^4 - 4\lambda^2h^2 + 2 = 0$$

or

$$\lambda^2h^2 = 0.59 = Ph^2/EI$$

Thus,

$$P_{cr} = \frac{2.36EI}{L^2} \qquad ◄$$

11.27

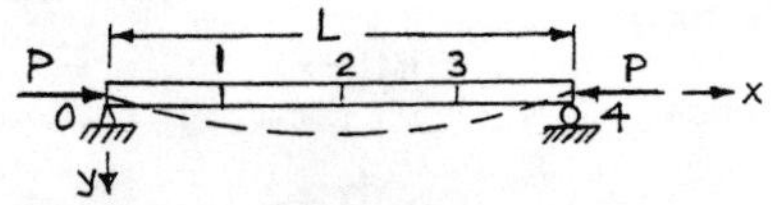

From symmetry: $v_1 = v_3$.
We have

$$I(x) = (1+\frac{2x}{L})EI_1 \qquad 0 \le x \le \frac{L}{2}$$

$$I(x) = (3-\frac{2x}{L})EI_1 \qquad \frac{L}{2} \le x \le L$$

Equation (11.24), gives at 0,1,2,3:

$$v_1 + v_{-1} = 0; \qquad v_1 = -v_{-1}$$

$$v_2 + \left[\frac{Ph^2}{3/2\,EI_1} - 2\right]v_1 = 0$$

(CONT.)

11.27 CONT.

$$v_1 + \left[\frac{Ph^2}{EI_1} - 2\right]v_2 + v_1 = 0$$

$$\left[\frac{Ph^2}{3/2\,EI_1} - 2\right]v_1 + v_2 = 0$$

The foregoing equations lead to

$$\begin{bmatrix} \frac{2}{3}\lambda^2h^2 - 2 & 1 \\ 2 & \frac{1}{2}\lambda^2h^2 - 2 \end{bmatrix} \begin{Bmatrix} v_1 \\ v_2 \end{Bmatrix} = 0$$

from which, since $v_1 \neq 0$ and $v_2 \neq 0$:

$$(\lambda^2h^2-6)(\lambda^2h^2-1) = 0$$

Thus,

$$P_{cr} = 16\frac{EI_1}{L^2} \qquad ◄$$

11.28

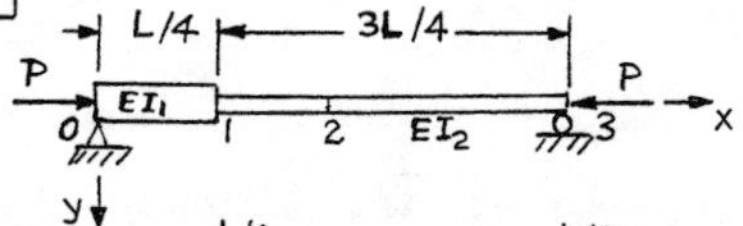

We have $\alpha_1 = \frac{L/4}{L/4} = 1 \qquad \alpha_2 = \frac{L/2}{L/4} = 2.$

Applying Eq. (11.25) at 1 and 2:

$$\left[\frac{PL^2}{16EI_1} - 2\right]v_1 + v_2 = 0 \qquad (a)$$

$$\frac{2}{3}v_1 + \left[\frac{PL^2}{16EI_1} - 1\right]v_2 = 0 \qquad (b)$$

Let

$$k_1 = \frac{PL^2}{16EI_1} \qquad k_2 = \frac{PL^2}{16EI_2}$$

Equations (a) and (b) yield then

$$\begin{vmatrix} k_1 - 2 & 1 \\ 2/3 & k_2 - 1 \end{vmatrix} =$$

from which

$$(k_1 - 2)(k_2 - 1) - \frac{2}{3} = 0$$

or

$$P^2\left[\frac{L^4}{256E^2I_1I_2}\right] - P\left[\frac{2L^2}{16EI_2} + \frac{L^2}{16EI_1}\right] + \frac{4}{3} = 0$$

Solution of this quadratic equation gives the critical load as follows:

$$P_{cr} = \frac{8E(2I_1+I_2)}{L^2} - \frac{8E}{L^2}[4I_1^2 + I_2^2 - 1.24I_1I_2]^{1/2}$$

In a special case, for $I_1 = I_2 = I$, the foregoing reduces to

$$P_{cr} = \frac{24EI}{L^2} - \frac{15.3\,EI}{L^2} = 8.7\,\frac{EI}{L^2} \qquad ◄$$

12.1

Components of stress are

$$\sigma_x = \frac{P}{A} + \frac{Mr}{I} = \frac{90}{\pi(0.05)^2} + \frac{3375(0.05)}{\pi(0.05)^4/4} = 45.84 \text{ MPa}$$

$$\tau_{xy} = \frac{Tr}{J} = \frac{4500(0.05)}{\pi(0.05)^4/2} = 22.92 \text{ MPa}$$

and $\sigma_y = \sigma_z = \tau_{xz} = \tau_{yz} = 0$

Thus, from Sec. 2.12 (in MPa):

$$\begin{bmatrix} \frac{2\sigma_x}{3} & \tau_{xy} & 0 \\ \tau_{xy} & -\frac{\sigma_x}{3} & 0 \\ 0 & 0 & -\frac{\sigma_x}{3} \end{bmatrix} \quad \begin{bmatrix} 30.56 & 22.92 & 0 \\ 22.92 & -15.28 & 0 \\ 0 & 0 & -15.28 \end{bmatrix}$$

12.2

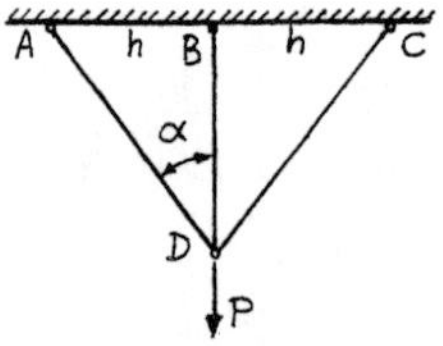

At instability, members AD (or DC) and BD become in length:

$$L'_{AD} = L'_{DC} = L_{AD} + \frac{n_1}{1-n_1} L_{AD} = \frac{L_{AD}}{1-n_1}$$

$$L'_{BD} = L_{BD} + \frac{n_2}{1-n_2} L_{BD} = \frac{L_{BD}}{1-n_2}$$

Therefore,

$$\left(\frac{L_{AD}}{1-n_1}\right)^2 = h^2 + \left(\frac{L_{BD}}{1-n_2}\right)^2 \qquad \text{(a)}$$

Initially, we have

$$L_{AD}^2 = h^2 + L_{BD}^2 \qquad \text{(b)}$$

Eliminating h from Eqs. (a) and (b), we obtain

$$\cos\alpha = \frac{L_{BD}}{L_{AD}} = \left(\frac{1-n_2}{1-n_1}\right)\sqrt{\frac{n_1(2-n_1)}{n_2(2-n_2)}}$$

For

$n_1 = 0.2$ and $n_2 = 0.3$

We have

$$\cos\alpha = \frac{0.7}{0.8}\left[\frac{0.2(1.8)}{0.3(1.7)}\right]^{1/2} = 0.7351$$

or

$$\alpha = 42.68^\circ$$ ◄

12.3

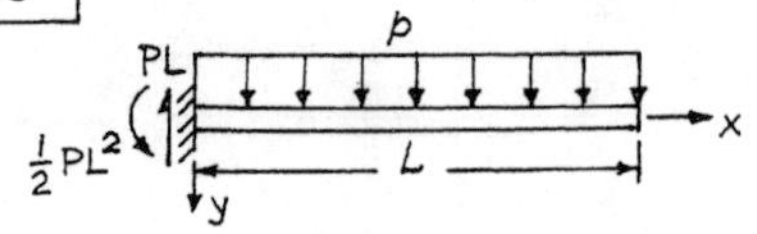

Expression for moment is

$$M(x) = \frac{1}{2}p(L-x)^2$$

Equation (12.7) is then

$$v'' = \left(\frac{M}{KI_n}\right)^{1/n} = \left(\frac{p}{2KI_n}\right)^{1/n}(L-x)^{2/n}$$

$$v' = -\frac{\lambda(L-x)^{2/n+1}}{(2/n+1)(2/n+2)} + c_1$$

or

$$v = \frac{\lambda(L-x)^{2/n+2}}{(2/n+1)(2/n+2)} + c_1x + c_2 \qquad \text{(a)}$$

Boundary conditions yield

$$v'(0)=0; \qquad c_1 = \frac{\lambda L^{2/n+1}}{2/n+1}$$

$$v(0)=0; \quad c_2 = -\frac{\lambda L^{2/n+2}}{(2/n+1)(2/n+2)}$$

Substituting these and Eq. (g) of Sec. 12.4 into Eq. (a), we obtain

$$v = \left(\frac{p}{2KI_n}\right)^{1/n}\left[\frac{(L-x)^{2/n+2}}{(2/n+1)(2/n+2)} - \frac{L^{2/n+2}}{(2/n+1)(2/n+2)} + \frac{L^{2/n+1}}{2/n+1}x\right]$$

For n=1, K=E, and x=L:

$$v = \frac{pL^4}{6EI} - \frac{pL^4}{24EI} = \frac{pL^4}{8EI}$$ ◄

12.4

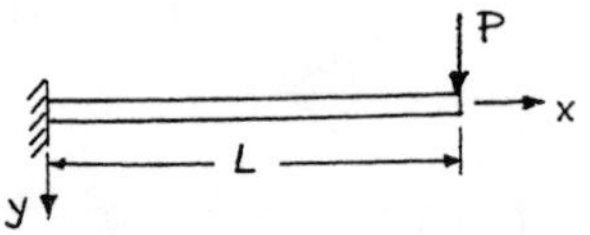

We have

$$M = P(L-x)$$

Hence,

$$v'' = \left(\frac{P}{KI_n}\right)^{1/n}(L-x)^{1/n} = \lambda(L-x)^{1/n}$$

$$v' = -\frac{\lambda(L-x)^{1/n+1}}{1/n+1} + c_1$$

$$v = \frac{\lambda(L-x)^{1/n+2}}{(1/n+1)(1/n+2)} + c_1x + c_2 \qquad \text{(a)}$$

Boundary conditions yield

$$v'(0)=0; \qquad c_1 = \frac{\lambda L^{1/n+1}}{1/n+1}$$

$$v(0)=0; \quad c_2 = -\frac{\lambda L^{1/n+2}}{(1/n+1)(1/n+2)}$$

(CONT.)

12.4 CONT.

Substituting these onto Eq. (a), we find an expression for the deflection.

For a special case of n=1 and K=E, the deflection at free end:

$$v(L) = \frac{PL^3}{2EI} - \frac{PL^3}{6EI} = \frac{PL^3}{3EI} \quad ◄$$

12.5

We have $\sigma = K\varepsilon^{1/4}$ and $I = \frac{2}{3}bh^3$

Here

$$\varepsilon = \frac{y}{h}\varepsilon_{max} = ay, \quad a = \varepsilon_{max}/h$$

Moment is thus,

$$M = \int \sigma\, y dA = \int_{-h}^{h} y\, K(ay)^{1/4}\, b dy$$

$$= \frac{8}{9} K a^{1/4} b h^{9/4}$$

But

$$\sigma_{max} = K\varepsilon_{max}^{1/4} = K a^{1/4} h^{1/4}$$

Hence, $M = (8/9)bh^2\sigma_{max}$, or

$$\sigma_{max} = 9M/8bh^2 = 3Mh/4I \quad ◄$$

12.6

$$M = -Px \qquad 0 \le x \le a$$

Hence,

$$v_1'' = (-\frac{Px}{KI_n})^{1/n} = -\lambda x^{1/n}$$

$$v_1' = -\frac{\lambda x^{1/n+1}}{1/n+1} + c_1$$

$$v_1 = -\frac{\lambda x^{1/n+2}}{(1/n+1)(1/n+2)} + c_1 x + c_2 \qquad (a)$$

Similarly, $M=-Pa$, at $a \le x \le (L-a)$:

$$v_2'' = (-\frac{Pa}{KI_n})^{1/n} = -\lambda a^{1/n}$$

$$v_2' = -\lambda a^{1/n} x + c_3$$

$$v_2 = -\lambda a^{1/n} \frac{x^2}{2} + c_3 x + c_4 \qquad (b)$$

Boundary conditions yield,

$$v_1(0) = 0; \qquad c_2 = 0$$

$$v_2(\tfrac{L}{2}) = 0; \qquad c_3 = \lambda a^{1/n} \frac{L}{2}$$

$$v_1'(a) = v_2'(a); \qquad c_1 = \lambda a^{1/n}\frac{L}{2} - \frac{\lambda a^{1/n+1}}{n(1/n+1)}$$

$$v_1(a) = v_2(a);$$

$$c_4 = \frac{\lambda a^{1/n+2}}{2} - \frac{\lambda a^{1/n+2}}{(1/n+1)(1/n+2)} - \frac{\lambda a^{1/n+2}}{n(1/n+1)}$$

(CONT.)

12.6 CONT.

Substitution of these constants into Eqs. (a) and (b) gives the required solution.

For a special case of n=1 and K=E, the midspan deflection is

$$v(\tfrac{L}{2}) = v_{max} = \frac{PaL^2}{8EI} - \frac{Pa^3}{6EI}$$

or

$$v_{max} = \frac{Pa}{24EI}(3L^2 - 4a^2) \quad ◄$$

We compute

$\bar{y} = 23.19$ mm (measured from top surface)

$I = 4.4(10^{-7})\ m^4$

and

$$v_{max} = \frac{0.45(8000)}{24(4.4\times10^{-7})200\times10^9}[3(1.2)^2 - 4(0.45)^2]$$

$$= 0.00598 \text{ m} = 6 \text{ mm} \quad ◄$$

12.7

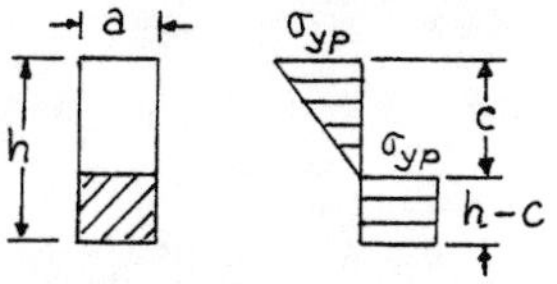

Referring to this figure,

$$(\sigma_{yp})ca/2 = (h-c)a\,\sigma_{yp}$$

or

$$c = 2h/3$$

Thus,

$$M = \frac{1}{2}\frac{2ha}{3}\sigma_{yp}\frac{2}{3}\frac{2h}{3} + \frac{h}{3}a\,\sigma_{yp}\frac{h}{6}$$

$$= (11/54)ah^2\sigma_{yp} \quad ◄$$

12.8

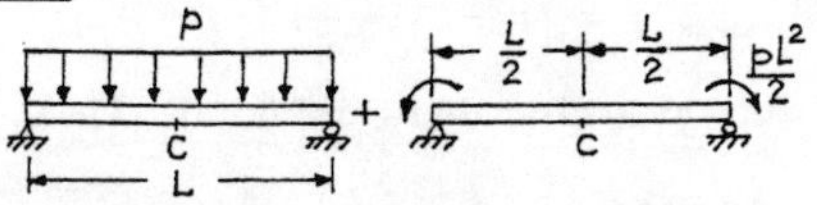

Deflection at C (Tables D.4 and 7.3), in <u>elastic range</u>:

$$v_{max} = \frac{5pL^4}{384EI} - \frac{(pL^2/12)L^2}{8EI} = \frac{pL^4}{384EI}$$

<u>Start of yielding</u>:

$$p = p_{yp} = 12M_{yp}/L^2$$

Thus,

$$v_{max} = \frac{M_{yp}L}{32EI} = \frac{(2bh^2\sigma_{yp}/3)L^2}{32E(2bh^3/3)}$$

$$= \sigma_{yp}L^2/32Eh \quad ◄$$

12.9

Initial yielding:

$$\sigma_{yp} = \frac{N_1}{\pi r^2} + \frac{4M_1}{\pi r^3} \qquad \text{(a)}$$

where

$$N_{yp} = \pi r^2 \sigma_{yp}, \quad M_{yp} = \sigma_{yp}\pi r^3/4$$

We express Eq. (a) in the form

$$\frac{N_1}{N_{yp}} + \frac{M_1}{M_{yp}} = 1 \qquad (1)$$

Fully plastic Deformation:

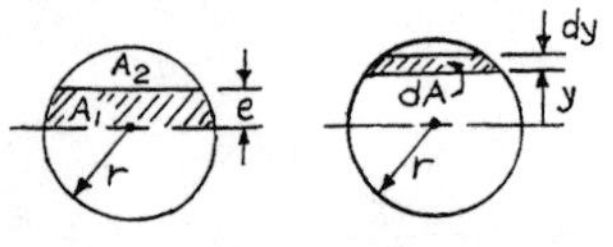

From a Mathematics Handbook table:

$$A_2 = \frac{\pi r^2}{2} - [e(r^2-e^2)^{1/2} + r^2 \sin^{-1}(\frac{e}{r})]$$

$$A_1 = e(r^2-e^2)^{1/2} + r^2 \sin^{-1}(\frac{e}{r})$$

Thus,

$$N_2 = 2\sigma_{yp}[e(r^2-e^2)^{1/2} + r^2 \sin^{-1}(\frac{e}{r})] \qquad (2)$$

We also write

$$dA = 2(r^2-y^2)^{1/2} dy$$

$$Q = \int_e^r 2y(r^2-y^2)^{1/2} dy = \frac{2}{3}(r^2-e^2)^{3/2}$$

where Q is the first moment of the area A_2. Hence,

$$M_2 = 2Q\sigma_{yp} = \frac{4}{3}(r^2-e^2)^{3/2}\sigma_{yp} \qquad (3)$$

$$M_u = \frac{4}{3} r^3 \sigma_{yp} = \frac{16}{3\pi} M_{yp}$$

Solving Eq. (3),

$$e = [r^2 - (\frac{3}{4}\frac{M_2}{\sigma_{yp}})^{2/3}]^{1/2}$$

Substituting this into Eq. (2):

$$N_2 = 2\sigma_{yp}\left[r^2 - (\frac{3}{4}\frac{M_2}{\sigma_{yp}})^{2/3}\right]^{1/2}(\frac{3}{4}\frac{M_2}{\sigma_{yp}})^3 +$$

$$2r^2\sigma_{yp}\sin^{-1}\left[1-(\frac{3}{4r^3}\frac{M_2}{\sigma_{yp}})^{2/3}\right]^{1/2}$$

The foregoing results in

$$\frac{\pi N_2}{2N_{yp}} = (\frac{3\pi}{16})^{1/3}(\frac{M_2}{M_{yp}})^{1/3}\left[1-(\frac{3\pi}{16})^{2/3}(\frac{M_2}{M_{yp}})^{2/3}\right]^{1/2}$$

$$+ \sin^{-1}\left[1-(\frac{3\pi}{16})^{2/3}(\frac{M_2}{M_{yp}})^{2/3}\right]^{1/2} \qquad (4)$$

(CONT.)

12.9 CONT.

The governing equations for yielding to impend and for fully plastic deformation aare given by Eqs. (1) and (4). A sketch of these, interaction curves , are shown below.

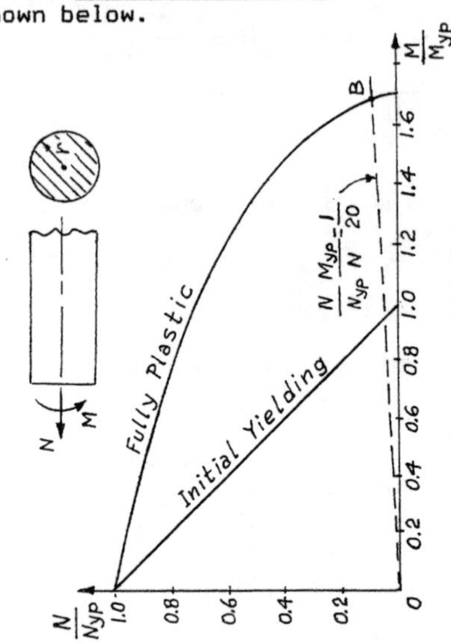

Figure P12.9

12.10

Let P=N. Then referring to Fig. P5.29, we have

$$M = Nd = N(0.05+0.05+0.025) = 0.125N$$

Thus,

$$N/M = 1/0.125$$

Also,

$$\frac{M_{yp}}{N_{yp}} = \frac{(\pi r^3/4)\sigma_{yp}}{\pi r^3 \sigma_{yp}} = \frac{0.025}{4}$$

and

$$M/M_{yp} = 20(N/N_{yp})$$

Now referring to Fig. P12.9, we find that

B(1.68, 0.084)

Therefore,

$$N_2 = 0.084N_{yp} = 0.084(\pi r^3 \sigma_{yp})$$

$$= 0.084\pi(0.025)^2(280\times10^6)$$

or

$$N = 46.18 \text{ kN}$$ ◄

12.11

$0 \le e \le h_1$ $\qquad$ $h_1 \le e \le h$

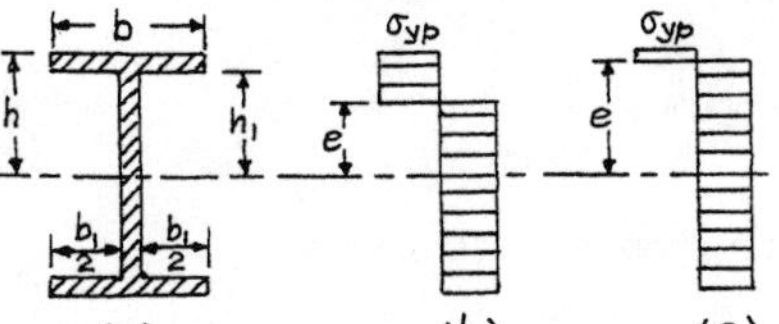

(a) $\qquad$ (b) $\qquad$ (c)

<u>Initial Yielding:</u>

$$\sigma_{yp} = N_1/A + M_1 c/I$$

$$= \frac{N_1}{2(bh - b_1 h_1)} + \frac{M_1 h}{2/3(bh^3 - b_1 h_1^3)} \qquad \text{(a)}$$

Here,

$$N_{yp} = 2(bh - b_1 h_1)\sigma_{yp}$$

$$M_{yp} = \frac{2}{3h}(bh^3 - b_1 h_1^3)\sigma_{yp}$$

Equation (a) may now be written

$$\frac{N_1}{N_{yp}} + \frac{M_1}{M_{yp}} = 1 \qquad (1)$$

<u>Fully-Plastic Deformation:</u>

<u>For $0 \le e \le h_1$ (Fig. b):</u>

$$N_2 = 2A_1\sigma_{yp} = 2(b - b_1)e\,\sigma_{yp}$$

from which

$$e = \frac{N_2}{2(b - b_1)\sigma_{yp}}$$

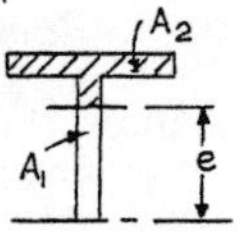

We have

$$A_2 = b(h - h_1) + (h_1 - e)(b - b_1)$$
$$= bh - eb + eb_1 - b_1 h_1$$

$$\bar{y} = \frac{\Sigma A y}{A_2} = \frac{b(h - h_1)\left(\frac{h - h_1}{2} + 1\right)}{A_2}$$

$$= \frac{1}{2}\,\frac{bh^2 - b_1 h_1^2 - e^2 b + e^2 b_1}{bh - eb + eb_1 - b_1 h_1}$$

Thus,

$$M_2 = 2\sigma_{yp} A_2 \bar{y}$$

$$= (bh^2 - b_1 h_1^2 - e^2 b^2 + e^2 b_1)\sigma_{yp} \qquad \text{(b)}$$

For $e = 0$: $M_2 = M_U = (bh^2 - b_1 h_1^2)\sigma_{yp}$. Substituting the given data, the preceding expressions become

$$N_{yp} = 1.68h^2\sigma_{yp}, \quad M_{yp} = 0.922h^3\sigma_{yp}$$
$$M_U = 1.113h^3\sigma_{yp}$$

and

$$M_U/M_{yp} = 1.21, \quad M_{yp} = 0.55hN_{yp}$$

Hence, Eq. (b) leads to

$$\frac{M_2}{1.21 M_{yp}} = 1 - \frac{1}{3.16}\left(\frac{N_2}{N_{yp}}\right)^2 \qquad \text{(2a)}$$

(CONT.)

12.11 CONT.

This is valid for

$0 \le e \le h_1$ $\quad$ or $\quad$ $0 \le \frac{N_2}{N_{yp}} \le 1.67$

<u>For $h_1 \le e \le h$ (Fig. c):</u>

$$N_2 = 2[0.7h + 0.2h + (e - 0.7h)2h]\,\sigma_{yp}$$

$$= (-2.52h^2 + 4eh)\sigma_{yp}$$

or

$$e = \frac{N_2}{4h\sigma_{yp}} + 0.63h$$

$$M_2 = 2h(h - e)2\left(e + \frac{h - e}{2}\right)\sigma_{yp}$$

$$= 2h^3\sigma_{yp} - 2he^2\sigma_{yp}$$

Hence,

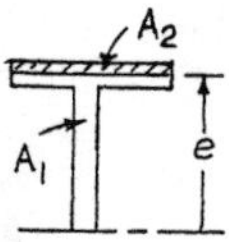

$$\frac{M_2}{2h^3\sigma_{yp}} = 1 - \frac{e^2}{h^2}$$

$$= 1 - \frac{N_2^2}{16h^4\sigma_{yp}^2} - \frac{0.315 N_2}{h^2\sigma_{yp}} - 0.397$$

or

$$\frac{1}{2.17}\frac{M_2}{M_{yp}} = 0.603 - 0.176\left(\frac{N_2}{N_{yp}}\right)^2$$
$$- 0.529\left(\frac{N_2}{N_{yp}}\right) \qquad \text{(2b)}$$

which is valid for

$h_1 \le e \le h$ $\quad$ or $\quad$ $1.67 \le \frac{N_2}{N_{yp}} \le 1$

Equations (1) and (2) are the governing expressions of the plastic bending. A sketch of these, interaction curves, are shown in in the figure given below.

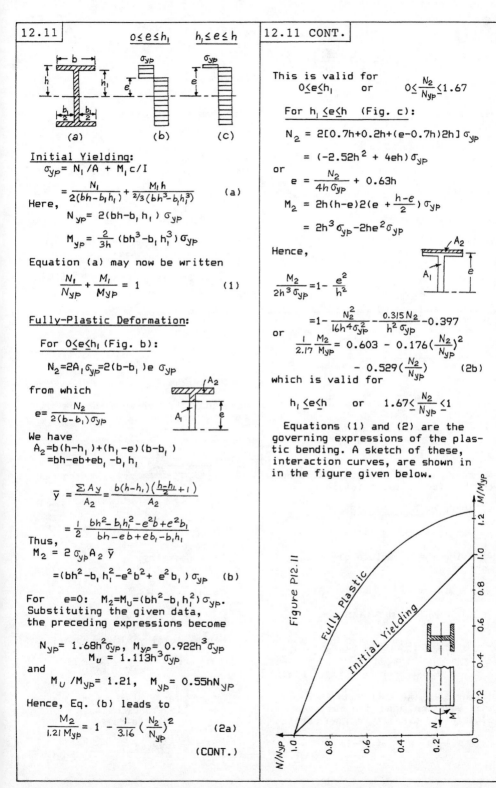

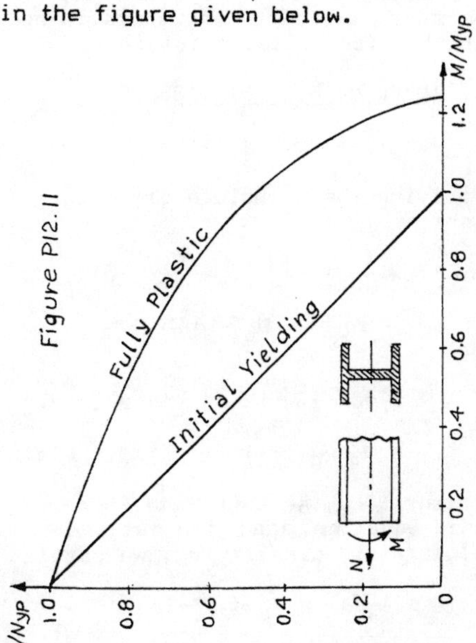

Figure P12.11

12.12

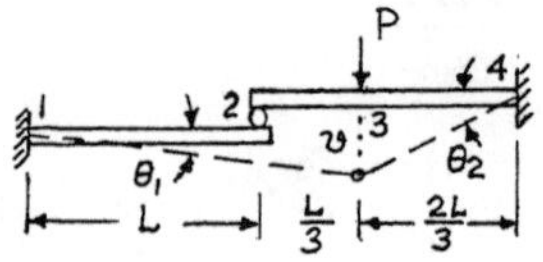

We have
$$v = \frac{4}{3}L\theta_1 = \frac{2}{3}L\theta_2$$
and
$$\delta v = \frac{4}{3}L\,\delta\theta_1 = \frac{2}{3}L\,\delta\theta_2$$
from which
$$2\,\delta\theta_1 = \delta\theta_2$$

Applying the principle of virtual work:
$$P_U\,\delta v = M_U\delta\theta_1 + M_U(\delta\theta_1 + \delta\theta_2) + M_U\,\delta\theta_2$$
or
$$\frac{4}{3}P_U L\,\delta\theta_1 = 6M\,\delta\theta_1$$

Solving,
$$P_U = 9M_U/2L$$ ◄

12.13

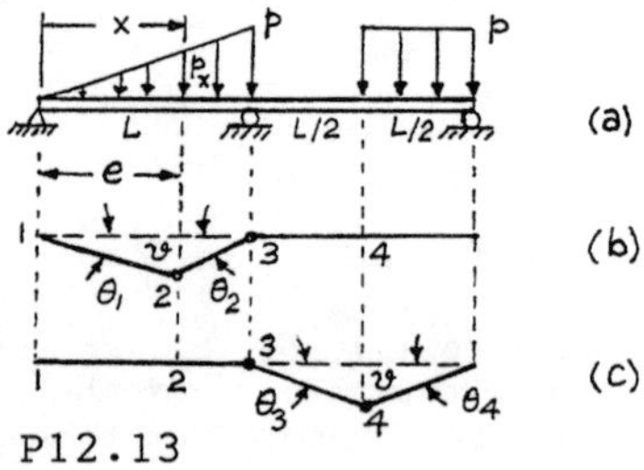

Fig. P12.13

We have two different modes to be checked for collapse loading.

Mode A, Fig. P12.13b:

$$\theta_1 = \theta_2(L-e)/e; \quad \theta_1 + \theta_2 = L\,\theta_2/e$$

Applying the principle of virtual work:

$$\delta\int_0^e (\theta_1 x)\frac{x}{L}p\,dx + \delta\int_e^L [\theta_1 e - \theta_2(x-e)]\frac{x}{L}p\,dx = M_U\,\delta\theta_1 + M_U\,\delta\theta_2 + M_U\,\delta\theta_2$$

or

$$\delta\int_0^e \theta_1 x\frac{x}{L}p\,dx + \delta\int_e^L [\theta_1 e - \theta_2(x-e)]\frac{x}{L}p\,dx = M_U\,\delta\theta_2\left(\frac{L}{e}+1\right) \qquad \text{(a)}$$

Integrating the left hand side of this equation, carrying out the algebra and simplifying, we obtain

$$\delta\theta_2\,p\left[\frac{L^2-e^2}{6}\right] = M_U\,\delta\theta_2\left(\frac{L}{e}+1\right)$$

(CONT.)

12.13 CONT.

Solving,
$$p = \frac{6M_U}{e(L-e)} \qquad \text{(b)}$$

Then, dp/de = 0 gives,
$$0 = -\frac{6M_U}{e^2(L-e)} + \frac{6M_U}{e(L-e)^2}; \qquad e = L/2$$

Introducing this value of e into Eq. (b), the collapse load is

$$P_U = 24M_U/L^2$$ ◄

Mode B, Fig. P12.13c:

From symmetry $\theta_3 = \theta_4 = \theta$. Principle of virtual work gives:

$$\delta\int_{L/2}^{L}\left[\theta\frac{L}{2} - \theta\left(x - \frac{L}{2}\right)\right]p\,dx = 3M_U\,\delta\theta$$

or

$$\delta\int_{L/2}^{L}[\theta L - \theta x]p\,dx = 3M_U\,\delta\theta$$

Integrating,

$$\delta\theta\frac{L^2 p}{8} = 3M_U\,\delta\theta$$

or

$$P_U = 24M_U/L$$ ◄

Note that the collapse load is the same for modes A and B.

12.14

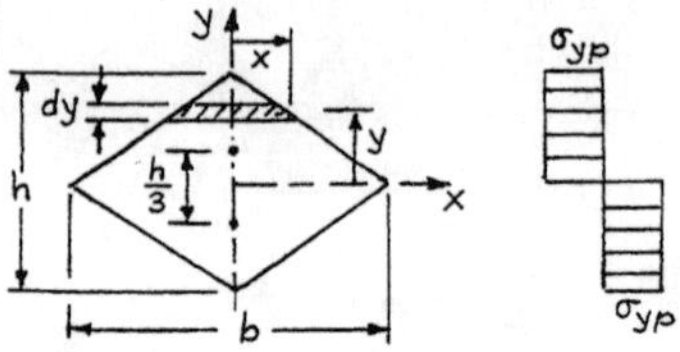

(a) We have $\quad M_{yp} = \sigma_{yp}I/h$

From geometry,
$$\frac{x}{b/2} = \frac{(h/2)-y}{h/2}; \qquad x = \frac{b(h-2y)}{2h}$$
Then,
$$I = 2\int_0^{h/2}(2x)dy(y^2) = \frac{bh^3}{48}$$

Hence, total yielding moment
$$M_{yp} = 2\frac{bh^3}{48}\frac{1}{h}\sigma_{yp} = \frac{bh^2}{24}\sigma_{yp}$$

Also,
$$M_U = \frac{1}{2}\text{(area of rhombus)} \times \text{(distance between centroids)}(\sigma_{yp})$$
$$= \frac{bh}{4}\left(\frac{h}{3}\right)\sigma_{yp} = \frac{bh^2}{12}\sigma_{yp}$$

(CONT.)

12.14 CONT.

Thus,

$$M_U/M_{yp} = 2$$

(b)

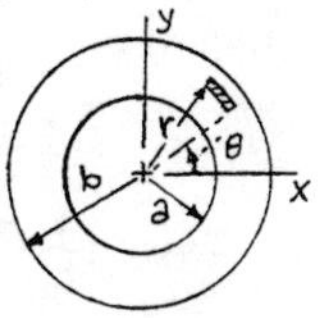

$$I = \pi(b^4-a^4)/4$$

$$M_{yp}=\frac{\pi}{4}(b^4-a^4)\frac{\sigma_{yp}}{b}=\frac{\pi}{4b}(b^4-a^4)\sigma_{yp}$$

Also, referring to the figure:

$$M_x = 2\int_0^{\pi}\int_0^{b} r^2\sin\theta \, dr \, d\theta$$

$$=\frac{4}{3}(b^3-a^3)$$

Hence,

$$y_c = \frac{M_x}{A}=\frac{4}{3}\,\frac{b^3-a^3}{\pi/2\,(b^2-a^2)}=\frac{4}{3\pi}\,\frac{b^3-a^3}{b^2-a^2}$$

We therefore have,

$$\frac{M_U}{M_{yp}}=\frac{\pi/2\,(b^2-a^2)\,\sigma_{yp}\,y_c}{(\pi/4b)(b^4-a^4)\,\sigma_{yp}}$$

$$=\frac{16b}{3\pi}\,\frac{b^3-a^3}{b^4-a^4}$$ ◄

12.15

For r=0, Eq. (f) of Sec. 12.8 leads to c_1 =0. Then, in <u>plastic zone</u>

$$\sigma_r = \sigma_{yp}-\rho\omega^2 r^2/3, \quad \sigma_\theta = \sigma_{yp}$$

If the plastic zone extends to radius c:

$$\sigma_c = \sigma_{yp}-\frac{\rho\omega^2 c^2}{3}$$

which may be found directly from Eq. (12.20) by setting a =0. The outer elastic zone is represented by an annular disk yielding at the inner radius c, wherein radial stress is σ_c.

We follow a procedure similar to that described in Sec. 12.8 for an annular disk. Boundary conditions:

$$(\sigma_r)_{r=c}=\sigma_c\,, \qquad (u)_{r=0}= 0$$

are substituted into Eqs. (8.27) to obtain c_1 and c_2. We then determine the stresses in the

(CONT.)

12.15 CONT.

elastic region as follows:

$$\sigma_r = \frac{\sigma_{yp}}{24N^2}\left[3(1+\nu)-(1+3\nu)\frac{c^4}{r^2b^2}\right](1-\frac{r^2}{b^2})$$

$$\sigma_\theta = \frac{\sigma_{yp}}{24N^2}\left[\frac{c^4}{b^4}(1+\frac{b^4}{r^4})(1+3\nu)+3\nu(3+\nu)-3(1+3\nu)\frac{r^2}{b^2}\right]$$

where

$$N^2= \frac{8+(1+3\nu)[(c/b)^2-1]}{24}$$ ◄

12.16

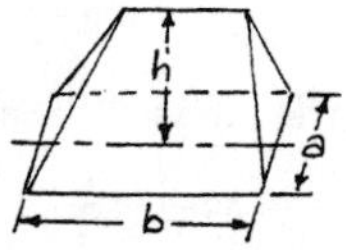

The sand volume is

$$V =\frac{1}{2}(b-a)ah + 2(\frac{1}{3}a^2\frac{h}{2})$$

Slope

$$\tau_{yp}= h/(a/2)$$

The ultimate torque is thus

$$T_U = \frac{(3b-a)\,a^2}{6}\,\tau_{yp}$$ ◄

The yield torque is given in Table 6.2, by letting $\tau=\tau_{yp}$.

12.17

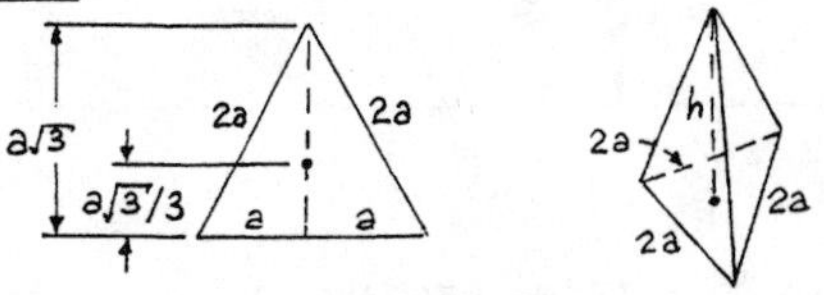

$$\text{Volume} =\frac{1}{3}h(\frac{1}{2}\cdot 2a\cdot a\sqrt{3})=\frac{\sqrt{3}}{3}ha^2$$

$$\text{Slope} =\frac{h}{a\sqrt{3}/3}=\frac{3h}{a\sqrt{3}}=\tau_{yp}, \quad h=\frac{a\sqrt{3}}{3}\tau_{yp}$$

(a) $$T_U = 2V = \frac{2}{3}a^3\tau_{yp}$$ ◄

(b) From Table 6.2:

$$\tau_A = 20\frac{T}{a_1^3}= \frac{20T}{8a^3}$$

Thus,

$$T_{yp} = \frac{8a^3}{20}\tau_{yp}$$

(c) Referring to the preceding results in items (a) and (b):

$$\frac{T_U}{T_{yp}} = \frac{5}{3}$$ ◄

12.18

Equilibrium condition, from Eq. (8.2):

$$\frac{d}{dr}(t\sigma_r) - \frac{t(\sigma_\theta-\sigma_r)}{r} = 0$$

Profile is, using Eq. (8.30) with s=1, $rt=at_a$. For full plasticity

$$\sigma_\theta - \sigma_r = \sigma_{yp}$$

Thus,

$$t\,\sigma_r = \int \frac{at_a}{r}\frac{\sigma_{yp}}{r}dr = -\frac{\sigma_{yp} a t_a}{r} + c_1$$

Since,

$$(\sigma_r)_{r=b} = 0; \quad c_1 = \frac{\sigma_{yp} a t_a}{b}$$

Then, noting that $\sigma_r = -p_i$ at $t=t_a$:

$$\frac{p_i}{\sigma_{yp}} = a\left(\frac{1}{r} - \frac{1}{b}\right)$$

or

$$p_i = \frac{a(b-r)}{rb}\sigma_{yp}$$ ◄

12.19

We have $\alpha = 1/2$ and substituting the given data into Eq. (12.32):

$$t_0 = \frac{0.606 p r_0}{(2K/\sqrt{3})(n/\sqrt{3})}$$

$$= \frac{0.606(14\times10^6)(0.5)}{(2\times900\times10^6/1.73)(0.2/1.73)^{0.2}}$$

$$= 0.0063 \text{ m} = 6.3 \text{ mm}$$ ◄

12.20

In this case, we have $\sigma_z > \sigma_\theta$. The total force is

$$P = 2\pi r t\,\sigma_1 \qquad (a)$$

where

$$\sigma_1 = \sigma_z \qquad \sigma_2 = \sigma_\theta$$

The values of r and t are given by Eqs. (g) and (f) of Sec. 12.9. Substituting these into Eq. (a):

$$P = 2\pi r_0 e^{-\varepsilon_1}\sigma_1$$

At instant of stability,

$$dP = (\partial P/\partial\sigma_1)d\sigma_1 + (\partial P/\partial\varepsilon_1)d\varepsilon_1 = 0$$

or

$$d\sigma_1/d\varepsilon_1 = \sigma_1$$

Equations (12.28) has the form

$$\sigma_1 = f(\alpha)\varepsilon_1^n$$

from which stability condition is

$$\varepsilon_1 = n$$

Then,

$$n = \left(\frac{\sigma_1}{K}\right)^{1/n}(\alpha^2-\alpha+1)\left(\frac{2-\alpha}{2}\right)$$

(CONT.)

12.20 CONT.

Solving,

$$\sigma_1 = K(2n)^n\left(\frac{1}{\alpha^2-\alpha+1}\right)^{(1-n)/2}\left(\frac{1}{2-\alpha}\right)^n \qquad (b)$$

Since $\varepsilon_1+\varepsilon_2+\varepsilon_3 = 0$, the maximum principal strain is

$$\varepsilon_1 = \ln\frac{L}{L_0} = \ln\frac{t}{t_0} + \ln\frac{r}{r_0}$$

Minimum strain is

$$\varepsilon_3 = \ln\frac{t}{t_0}\ ; \qquad t = t_0 \ln^{-1}\varepsilon_3$$

Thus,

$$\sigma_1 = \frac{P}{2\pi rt} = \frac{P}{2\pi r_0 \ln^{-1}\varepsilon_2\, t_0 \ln^{-1}\varepsilon_3}$$

Substituting Eqs. (12.28b) and (12.28c) into this equation:

$$\sigma_1 = \frac{P}{2\pi r_0 t_0 \ln^{-1}\{[(\frac{\sigma_1}{K})^{1/n}(\alpha^2-\alpha+1)\frac{1-n}{2}(\frac{\alpha}{2}-1)]\}}$$

or

$$t_0 = \frac{P}{2\pi r_0\sigma_1 \ln^{-1}\{[(\frac{\sigma_1}{K})^{1/n}(\alpha^2-\alpha+1)\frac{1-n}{2}(\frac{\alpha}{2}-1)]\}}$$ ◄

where, σ_1 is given by Eq. (b).

Results of the preceding expressions are simplified by setting $\alpha=1/2$.

12.21

Since the mean radius does not change, we have $d\varepsilon_\theta = 0$. We also take $\sigma_3 = \sigma_r = 0$. Material is incompressible $d\varepsilon_L = d\varepsilon_3$, or $\varepsilon_L = -\varepsilon_3$.

The Levy-Mises equation is thus

$$\frac{d\varepsilon_L}{\sigma_L-\sigma_\theta} = \frac{d\varepsilon_L}{\sigma_\theta}$$

or

$$\sigma_L = 2\sigma_\theta$$

Hence,

$$\bar\sigma = \frac{\sqrt{3}}{2}\sigma_L$$

and

$$\bar\varepsilon = (2/3)d\varepsilon_L = -(2/\sqrt{3})d\varepsilon_3$$

Radius r_0 = constant. Therefore,

$$P = 2\pi r_0 \cdot t\,\sigma_L$$

At instant of instability, dP = 0:

$$\frac{d\sigma_L}{\sigma_L} = -\frac{dt}{t} = -d\varepsilon_3 = d\varepsilon_L$$

Hence,

$$\bar\varepsilon = \frac{2}{\sqrt{3}}n = \frac{2}{\sqrt{3}}\varepsilon_L$$

$$\bar\sigma = K\left(\frac{2}{\sqrt{3}}\right)^n = \frac{\sqrt{3}}{2}\sigma_L, \qquad t = t_0 e^{-n}$$

At instant of instability, the load is then given by Eq.(P12.21).

12.22

(a) Using Eq. (8.11):

$$p_{yp} = \frac{420}{3}\frac{36-25}{2(36)} = 21.39 \text{ MPa}$$

We have k=1, and thus,

$$p_u = \frac{\sigma_{yp}}{f_s}\ln\frac{b}{a} = \frac{420}{3}\ln\frac{6}{5}$$

$$= 25.53 \text{ MPa} \quad ◄$$

(b)

$$\left(\frac{\sigma_{yp}}{f_s}\right)^2 = \sigma_\theta^2 - \sigma_\theta\sigma_r + \sigma_r^2$$

or

$$(140)^2 = p_{yp}^2\left[\left(\frac{36+25}{36-25}\right)^2 + \frac{36+25}{36-25} + 1\right]$$

Solving,

$$p_{yp} = 22.92 \text{ MPa} \quad ◄$$

Now $k = 2/\sqrt{3}$, and hence,

$$p_u = \frac{2}{\sqrt{3}}(25.53) = 29.48 \text{ MPa} \quad ◄$$

12.23

Using Eqs. (12.47) and (12.48) with k=1 and r=0.25 m,

$$p_u = 400\ln(0.3/0.2) = 162.2 \text{ MPa}$$

$$\sigma_r = -400\ln(0.3/0.25) = -72.93 \text{ MPa}$$

and

$$\sigma_\theta = 400[1-\ln\frac{0.3}{0.25}] = 327.1 \text{ MPa}$$

$$\sigma_z = \frac{1}{2}(\sigma_r + \sigma_\theta) = 127.1 \text{ MPa}$$

Unloading from p_u. At r=0.25 m, Eqs. (8.12) and (8.20):

$$\sigma_r = \frac{0.2^2(162.2)}{0.3^2-0.2^2}\left(1-\frac{0.3^2}{0.25^2}\right)$$

$$= -57.09 \text{ MPa}$$

$$\sigma_\theta = \frac{0.2^2(162.2)}{0.3^2-0.2^2}\left(1+\frac{0.3^2}{0.25^2}\right) \quad ◄$$

$$= 316.6 \text{ MPa}$$

$$\sigma_z = 162.2\,\frac{0.2^2}{0.3^2-0.2^2} = 129.8 \text{ MPa}$$

Residual stresses at r=0.25 m:

$$(\sigma_\theta)_{res.} = 327.1-316.6 = 10.5 \text{ MPa}$$

$$(\sigma_r)_{res.} = -72.93+57.09 = -15.84 \text{ MPa} \quad ◄$$

$$(\sigma_z)_{res.} = 127.1-129.8 = -2.7 \text{ MPa}$$

12.24

(a) Equation (12.50):

$$p_c = k\sigma_{yp}\frac{3^2-2^2}{2(3)^2} = 0.2778k\,\sigma_{yp} \quad ◄$$

(b) Equation (12.51a):

$$(\sigma_r)_{r=a} = k\sigma_{yp}[\ln\frac{1}{2} - 0.2778]$$

$$= -0.9709\; k\,\sigma_{yp} \quad ◄$$

(c) Equation (12.49b):

$$(\sigma_\theta)_{r=b} = \frac{0.2778k\,\sigma_{yp}(3^2)}{3^2-2^2}\left(1+\frac{3^2}{3^2}\right)$$

$$= 0.4444\; k\,\sigma_{yp} \quad ◄$$

Equation (12.49b):

$$(\sigma_\theta)_{r=c} = \frac{0.2778k\sigma_{yp}(3^2)}{3^2-2^2}\left(1+\frac{3^2}{2^2}\right)$$

$$= 0.7222\; k\,\sigma_{yp}$$

Equation (12.51b):

$$(\sigma_\theta)_{r=a} = k\,\sigma_{yp}(1+\ln\frac{1}{2}-0.2778)$$

$$= 0.0291\; k\,\sigma_{yp} \quad ◄$$

We see from these results that the maximum stress occurs at the elastic-plastic boundary.

13.1

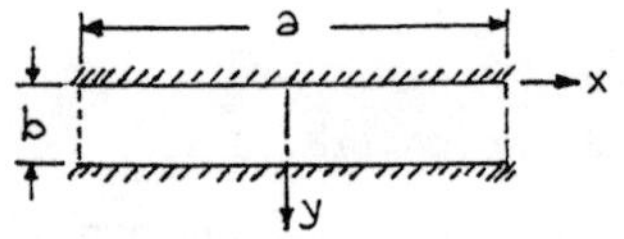

(a)

Boundary conditions at $y = 0$ and $y=b$:

$$w = 0 \qquad dw/dy = 0 \qquad \text{(a)}$$

We have

$$w^{IV} = p_o/D, \quad w'' = \frac{p_o y^2}{2D} + c_1 y + c_2$$

$$w' = \frac{p_o y^3}{6D} + \frac{1}{2}c_1 y^2 + c_2 y + c_3$$

$$w = \frac{p_o y^4}{24D} + \frac{c_1 y^3}{6} + \frac{c_2 y^2}{2} + c_3 y + c_4$$

Conditions (a) yield $c_4=0$, $c_3=0$,

$$c_1 = -\frac{p_o b}{2D} \qquad c_2 = \frac{p_o b^2}{12D}$$

Thus,

$$w = \frac{p_o b^4}{24D}\left[\left(\frac{y}{b}\right)^4 - 2\left(\frac{y}{b}\right)^3 + \left(\frac{y}{b}\right)^2\right] \quad ◄$$

(b)

Integrating twice, we have

$$w'' = \frac{p_o b^4}{24D}\left[\frac{12}{b^4}y^2 - \frac{12}{b^3}y + \frac{2}{b^2}\right]$$

For $y = b/2$:

$$d^2w/dy^2 = -p_o b^2/24D$$

Hence,

$$M_y = p_o b^4/24$$

Thus,

$$\sigma_{y,max} = \frac{p_o}{4}\left(\frac{b}{t}\right)^2, \quad \sigma_{x,max} = \frac{p_o}{12}\left(\frac{b}{t}\right)^2$$

Similarly, for $y = 0$:

$$d^2w/dy^2 = -\frac{p_o b^2}{12D}, \qquad M_y = \frac{p_o b^2}{12}$$

and

$$\sigma_{y,max} = \frac{1}{2}p_o\left(\frac{b}{t}\right)^2 = \sigma_{max}$$

$$\sigma_{x,max} = \frac{1}{6}p_o\left(\frac{b}{t}\right)^2 \quad ◄$$

13.2

We have $r_x = r$, $r_y = \infty$ ($K_y = 0$) and

$$r = 0.12 \text{ m}, \quad t = 0.3(10^{-3}) \text{ m}$$

Equation (13.3b):

$$\varepsilon_{max} = \frac{t}{2r} = \frac{0.3}{2(120)} = 1250\,\mu$$

Equation (13.5):

$$\sigma_{max} = -\frac{Et}{2(1-\nu^2)r} = -\frac{200\times10^9(0.3)}{2(0.91)\,120}$$

$$= 274.7 \text{ MPa} \quad ◄$$

13.3

Using Eq. (13.8),

$$D = \frac{9(200\times10^9)(0.012)^3}{12(8)} = 32{,}400$$

From Example 13.1:

$$w = \left(\frac{b}{\pi}\right)^4 \frac{p_o}{D}\sin\left(\frac{\pi y}{b}\right)$$

$$w_{max} = \left(\frac{0.6}{\pi}\right)^4 \frac{20\times10^3}{32{,}400} = 0.82(10^{-3}) \text{ m}$$

$$= 0.82 \text{ mm} \quad ◄$$

$$M_y = -D\frac{d^2w}{dy^2} = -\left(\frac{b}{\pi}\right)^2 p_o \sin\frac{\pi y}{b}$$

Thus,

$$\sigma_{y,max} = 6M_{y,max}/t^2 = 0.6p_o\left(\frac{b}{t}\right)^2$$

$$= 0.6p_o(20\times10^3)\left(\frac{0.6}{0.012}\right)^2$$

$$= 30 \text{ MPa}$$

$$\sigma_{x,max} = \nu(30) = 10 \text{ MPa}$$

Then,

$$\varepsilon_{y,max} = \frac{1}{E}(\sigma_{y,max} - \nu\sigma_{x,max})$$

$$= \frac{1}{200\times10^3}\left(30 - \frac{10}{3}\right) = 133\,\mu \quad ◄$$

and

$$r_y = \frac{t}{2\varepsilon_{y,max}} = \frac{0.012\times10^6}{2(133)} = 45.1 \text{ m} \quad ◄$$

13.4

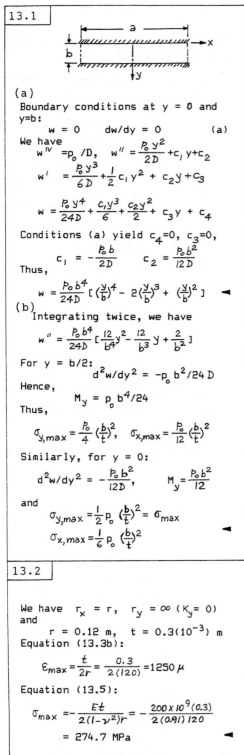

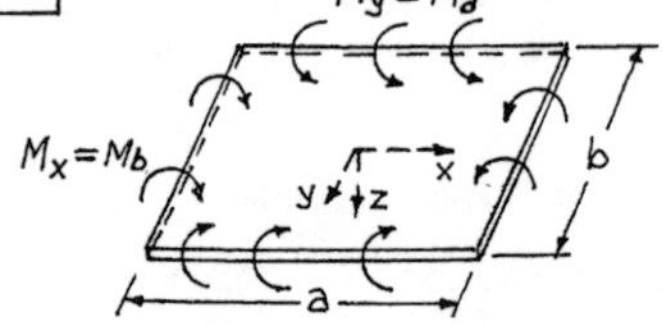

(a)

Using Eq. (13.7),

$$\frac{\partial^2 w}{\partial x^2} = -\frac{M_b - \nu M_a}{D(1-\nu^2)}$$

$$\frac{\partial^2 w}{\partial y^2} = -\frac{M_a - \nu M_b}{D(1-\nu^2)}, \qquad \frac{\partial^2 w}{\partial x \partial y} = 0 \qquad \text{(a)}$$

Integrating these equations,

$$w = -\frac{M_b - \nu M_a}{2D(1-\nu^2)}x^2 - \frac{M_a - \nu M_b}{2D(1-\nu^2)}y^2 + c_1 x + c_2 y + c_3$$

If the origin of xyz is located at the center and midplane of the plate, the c's will vanish, and

(CONT.)

13.4 CONT.

$$w = -\frac{M_b - \nu M_a}{2D(1-\nu^2)}x^2 - \frac{M_a - \nu M_b}{2D(1-\nu^2)}y^2 \qquad (b)$$

(b) By setting $M_a = -M_b$ in Eq. (a):

$$\frac{\partial^2 w}{\partial x^2} = -\frac{\partial^2 w}{\partial y^2} = \frac{M_a}{D(1-\nu)} = \frac{1}{r_x} = -\frac{1}{r_y}$$

Integrating and locating the origin of xyz, as in item (a):

$$w = \frac{M_a}{2D(1-\nu)}(x^2 - y^2) \qquad ◄$$

13.5

Equation (13.19) becomes,

$$p_{mn} = \frac{144P}{a^4b^4}\int_0^a\left[\int_0^b (b-y)\sin\frac{n\pi y}{b}dy\right] \cdot (a-x)\sin\frac{m\pi x}{a}dx$$

Integrating by parts,

$$p_{mn} = \frac{144P}{a^4b^4}\,\frac{b^2}{n\pi}\,\frac{a^2}{m\pi} = \frac{144P}{mn\pi^2a^2b^2} \qquad (a)$$

Substituting Eqs. (13.18) into $\nabla^4 w = p/D$:

$$a_{mn} = \frac{p_{mn}}{\pi^4 D(m^2/a^2 + n^2/b^2)^2} \qquad (b)$$

Inserting Eqs. (a) and (b) into Eq. (13.18b), we find the required expression for the deflection.

13.6

(a) From Eq. (13.19), we obtain

$$p_{mn} = p_o$$

Then, Eq. (13.20) becomes for a square plate (a=b):

$$w = \frac{p_o a^4}{\pi^4 D}\sum\sum \frac{\sin\frac{m\pi x}{a}\sin\frac{n\pi y}{b}}{(m^2+n^2)^2} \qquad (a)$$

At x = y = a/2,

$$w_{max} = \frac{p_o a^4}{\pi^4 D}\sum\sum \frac{(-1)^{\frac{m+n}{2}-1}}{(m^2+n^2)^2} \qquad ◄$$

(b)

$$D = \frac{210\times10^9(0.025)^3}{12(1-0.3^2)} = 300{,}480.77$$

Equation (a) is then,

$$8(10^{-3}) = \frac{p_o(3)^4}{\pi^4(300{,}480.77)}\left[\frac{1}{4} - \frac{1}{100}\right]$$

or

$$p_o = 12.05 \text{ kPa} \qquad ◄$$

13.7

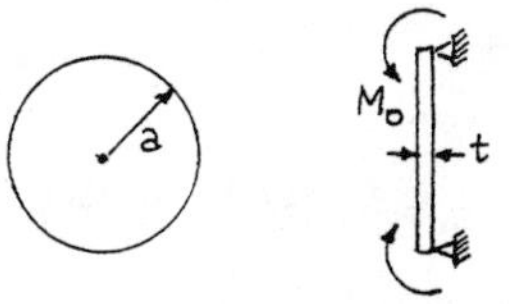

Since Q = 0, Eq. (13.24c) becomes

$$\frac{d}{dr}\left[\frac{1}{r}\frac{d}{dr}\left(r\frac{dw}{dr}\right)\right] = 0$$

from which, after integration,

$$w = -\frac{c_1 r^2}{4} - c_2 \ln\frac{r}{a} + c_3 \qquad (a)$$

Substituting this into Eq. (13.24a):

$$M_r = D\left[\frac{c_1}{2} - \frac{c_2}{r^2} + \nu\left(\frac{c_1}{2} + \frac{c_2}{r^2}\right)\right] \qquad (b)$$

Boundary conditions

$(M_r)_{r=a} = M_o \qquad w(a) = 0$

$(dw/dr)_{r=0} = 0$

yield $c_2 = 0$ and

$$c_1 = \frac{2M_o}{D(1+\nu)} \qquad c_3 = \frac{M_o a^2}{2D(1+\nu)}$$

Equation (a) becomes then,

$$w = \frac{M_o}{2D(1+\nu)}(a^2 - r^2) \qquad ◄$$

Introducing this into Eqs. (13.24a) and (13.24b) yields $M_r = M_\theta = M_o$. Hence,

$$\sigma_{r,max} = \sigma_{\theta,max} = \frac{6M_o}{t^2}$$

13.8

We have Q_r=0 and Eq. (13.24c) after integration gives

$$w = -\frac{c_1 r^2}{4} - c_2 \ln\frac{r}{a} + c_3 \qquad (a)$$

Introducing this into Eq.(13.24a):

$$M_r = D\left[\frac{c_1}{2} - \frac{c_2}{r^2} + \nu\left(\frac{c_1}{2} + \frac{c_2}{r^2}\right)\right] \qquad (b)$$

Boundary conditions

$w(a) = 0 \qquad M_r(b) = M_o$

and Eqs. (a) and (b) result in

$$c_1 = \frac{2b^2M_o}{(1+\nu)(b^2-a^2)}, \qquad c_2 = \frac{a^2b^2M_o}{(1-\nu)(b^2-a^2)}$$

$$c_3 = \frac{a^2b^2M_o}{2(1+\nu)D(a^2-b^2)}$$

Carrying these into Eq. (a), we have the equation for deflection.

13.9

From Example 13.3:

$$\sigma_1 = \sigma_{r,max} = \frac{3}{4}p_o\left(\frac{a}{t}\right)^2, \quad \sigma_2 = -\frac{1}{4}p_o\left(\frac{a}{t}\right)^2$$

Maximum shearing stress is then

$$\tau_{max} = \frac{1}{2}(\sigma_1 - \sigma_2) = p_o\left(\frac{a}{t}\right)^2$$

According to the maximum shear stress theory:

$$\frac{\sigma_{yp}}{2f_s} = p_o\left(\frac{a}{t}\right)^2$$

Introducing the given data,

$$\frac{100(10^6)}{2f_s} = (0.4\times10^6)\left(\frac{0.2}{0.02}\right)^2$$

or

$$f_s = 1.25 \quad ◄$$

13.10

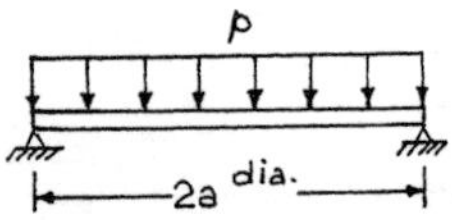

Solution proceeds as in Example 13.3. Boundary conditions are

$$(w)_{r=a} = 0 \qquad (M_r)_{r=a} = 0 \qquad (a)$$

We have

$$\frac{D}{p}w = \frac{r^4}{64} - \frac{c_2 r^2}{4} + c_4$$

Carrying Eqs. (a) and (b) into Eq. (13.24a), we obtain two equations. From these c_2 and c_4 are evaluated. In so doing, and substituting the values obtained into Eq. (b):

$$w = \frac{p(a^2-r^2)}{64D}\left(\frac{5+\nu}{1+\nu}a^2 - r^2\right) \quad ◄$$

13.11

Referring to Example 13.2:

$$\sigma_{max} = \frac{6M_{max}}{t^2} = 6(0.0534p_o)\left(\frac{50}{2}\right)^2$$

Thus, $\quad 240(10^6) = 200.25p_o$

or

$$p_o = 1.2 \text{ MPa} \quad ◄$$

Similarly,

$$w_{max} = 0.0454p_o\frac{a^4}{Et^3}$$

$$= 0.0454(1.2\times10^6)\frac{(0.5)^4}{70\times10^9(0.02)^3}$$

$$= 0.0194 \text{ m} = 19.4 \text{ mm} \quad ◄$$

13.12

We observe that

$$\int_0^a \sin\frac{m\pi x}{a}\sin\frac{m'\pi x}{a}dx = \int_0^b \sin\frac{n\pi y}{b}\sin\frac{n'\pi y}{b}dy = 0$$

if $m \neq m'$ and $n \neq n'$. Therefore, integrating, we consider only the squares of the terms in the paranthesis in Eq. (b) of Sec. 13.7. Using the formula:

$$\int_0^a\int_0^b \sin^2\frac{m\pi x}{a}\sin^2\frac{n\pi y}{b}dxdy = \frac{ab}{4}$$

calculation of the first term of the integral in Eq. (b) gives

$$\frac{\pi^4 abD}{8}\sum_m^\infty\sum_n^\infty a_{mn}^2\left(\frac{m^2}{a^2}+\frac{n^2}{b^2}\right)^2 \quad ◄$$

Also, the second term of the integral in Eq. (b):

$$p_o\int_0^a\int_0^b a_{mn}\sin\frac{m\pi x}{a}\sin\frac{n\pi y}{b}dxdy$$

$$= \frac{p_o ab}{\pi^2 mn}a_{mn}(1-\cos m\pi)(1-\cos n\pi)$$

$$= \frac{4p_o ab}{\pi^2 mn}a_{mn} \qquad (m,n = 1,3,\ldots) \quad ◄$$

13.13

Deflection is given by Eq.(13.18b). Loading is expressed as follows:

$$p = 2p_o\frac{x}{a} \qquad 0 \le x \le \frac{a}{2}$$

$$p = 2p_o - 2p_o\frac{x}{a} \qquad \frac{a}{2} \le x \le a \qquad (a)$$

Potential energy, Eq. (13.33):

$$W = 2\sum\sum\int_0^{a/2}\int_0^a \frac{2p_o x}{a}\sin\frac{m\pi x}{a}\sin\frac{n\pi y}{a}dxdy$$

$$= \sum\sum\frac{8p_o a^2}{m^2 n\pi^3}a_{mn}\sin\frac{m\pi}{2} \qquad (b)$$

Strain energy, Eq. (13.32):

$$U = \frac{D}{2}\int_0^a\int_0^a\sum\sum\left[a_{mn}\left(\frac{m^2\pi^2}{a^2}+\frac{n^2\pi^2}{a^2}\right)\sin\frac{m\pi x}{a}\sin\frac{n\pi y}{a}\right]^2 dxdy$$

$$= \frac{1}{8}D\pi^4 a^2\sum\sum a_{mn}^2\left(\frac{m^2}{a^2}+\frac{n^2}{a^2}\right)^2$$

We thus have

$$\frac{\partial\Pi}{\partial a_{mn}} = \frac{D\pi^4 a^2}{4}a_{mn}\left(\frac{m^2}{a^2}+\frac{n^2}{a^2}\right)^2 - \frac{8p_o a^2}{m^2 n\pi^3}\sin\frac{m\pi}{2} = 0$$

or

$$a_{mn} = \frac{32p_o a^4\sin(m\pi/2)}{m^2 n\pi^7 D(m^2+n^2)^2} \qquad (c)$$

$(m,n = 1,3,.)$. Substitute this into Eq. (13.18b) to obtain deflection.

13.14

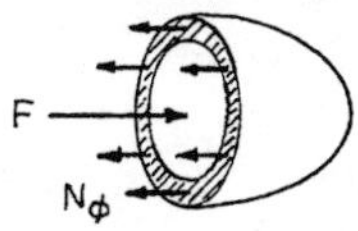

We have

$$r_\phi = 180 - 1 = 179 \text{ mm}$$
$$r_\theta = 80 - 1 = 79 \text{ mm}$$
$$p = -0.08 \text{ MPa} \qquad F = \pi r_i^2 p$$

Equation (13.46b) gives

$$N_\phi = \frac{F}{2\pi r_\theta (1)} = \frac{\pi (0.078)^2 (0.08\times 10^6)}{2\pi (0.079)}$$
$$= 3.081 \text{ kN}$$

and

$$\sigma_\phi = \frac{3081}{0.002} = 1.541 \text{ MPa} \quad ◄$$

Using Eq. (13.46a),

$$\frac{\sigma_\theta}{0.079} + \frac{1.541}{0.179} = \frac{0.08}{0.02}$$

or

$$\sigma_\theta = 2.48 \text{ MPa} = \sigma_{max} \quad ◄$$

13.15

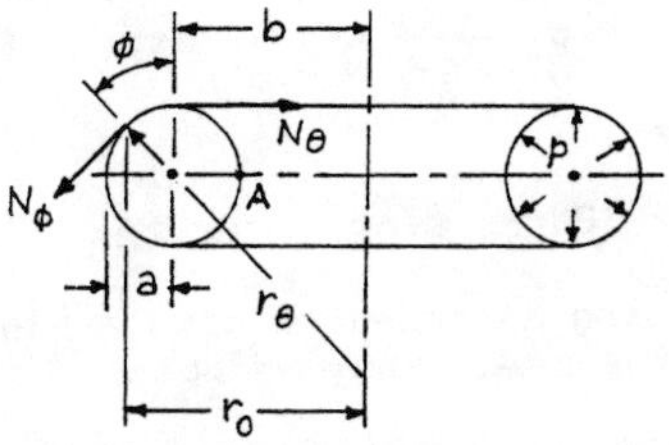

Consider the portion of shell defined by ϕ . Vertical equilibrium of forces yields,

$$2\pi r_0 N_\phi \sin\phi = \pi p (r_0^2 - b^2)$$

from which

$$N_\phi = \frac{p(r_0^2 - b^2)}{2 r_0 \sin\phi} = \frac{pa(r_0+b)}{2r_0}$$

or

$$N_\phi = \frac{pa}{b + a\sin\phi}\left(\frac{a}{2}\sin\phi + b\right) \quad ◄$$

Substituting N_ϕ into Eq. (13.46a), setting $p_z = -p$, and $r_\phi = a$:

$$N_\theta = \frac{p r_\theta (r_0 - b)}{2r_0} = \frac{pa}{2} \quad ◄$$

Since $\sin\phi = r_0/r_\theta = (r_0 - b)/a$, from symmetry.

Note that N_θ is constant througout the shell from the condition of symmetry.

13.16

Referring to Solution of Prob. 13.15, we have

$$2a = (1-0.7)/2, \qquad a = 0.075 \text{ m}$$
$$2b = (1+0.7)/2, \qquad b = 0.425 \text{ m}$$

At point A (crotch):

$$\sigma_A = \sigma_{\phi,max} = pa(r_0 + b)/2r_0 t$$

or

$$t = pa(r_0 + b)/2r_0 \sigma_{\phi,max}$$
$$= \frac{2(10^6)(0.075)(0.35+0.425)}{2(0.35)(210\times 10^6)}$$
$$= 0.791 \text{ mm} = t_{req.} \quad ◄$$

Similarly,

$$\sigma_\theta = pa/2t$$

or

$$t = \frac{2(10^6)\,75}{2(210\times 10^6)} = 0.357 \text{ mm}$$

13.17

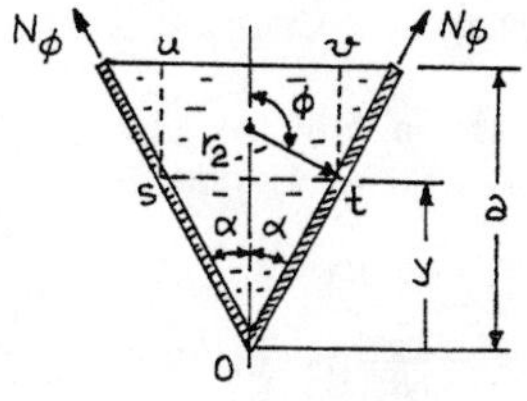

Pressure at any level st is

$$p_r = \gamma (a-y)$$

We have

$$\phi = \frac{\pi}{2} + \alpha \qquad r_0 = y\tan\alpha$$

Thus, the first of Eqs. (13.49) becomes

$$N_\theta = \frac{\gamma (a-y) y \tan\alpha}{\cos\alpha}$$

and

$$\sigma_\theta = \frac{\gamma (a-y) y}{t} \frac{\tan\alpha}{\cos\alpha} \quad ◄$$

The load F is equal to the weight of liquid in cylindrical portion stuv:

$$F = -\pi \gamma y^2 (a - y + \tfrac{y}{3}) \tan^2\alpha$$

Then second of Eqs. (13.49) gives

$$N_\phi = \gamma y (a - \tfrac{2}{3}y) \tan\alpha / 2\cos\alpha$$

and

$$\sigma_\phi = \frac{\gamma y (a - 2y/3)}{2t} \frac{\tan\alpha}{\cos\alpha} \quad ◄$$

13.18

Expressions for the components of pressure are:

$$p_\theta = -p\cos\theta \qquad p_r = p\sin\theta \qquad p_x = 0$$

Thus,

$$N_{x\theta} = -\int[-p\cos\theta + \frac{1}{a}(-pa\cos\theta)]dx + f_1(\theta)$$

$$= 2px\cos\theta + f_1(\theta) \qquad (a)$$

$$N_\theta = -pa\sin\theta \qquad (b)$$

$$N_x = -\int\frac{2}{a}(-px\sin\theta)dx + \frac{df_1}{d\theta} + f_2(\theta)$$

$$= \frac{1}{a}px^2\sin\theta + x\frac{df_1}{d\theta} + f_2(\theta) \qquad (c)$$

Boundary conditions,

$N_x = 0$ at $x=0$ and $x=L$

give $f_2 = 0$ and

$$\frac{pL^2}{a}\sin\theta + L\frac{df_1}{d\theta} = 0$$

or

$$f_1(\theta) = -\frac{pL}{2}\sin\theta + c$$

Note that no torque is applied to the shell; c=0. Hence,

$$N_\theta = -pa\sin\theta$$

$$N_x = -\frac{L-x}{a}px\sin\theta \quad ◄$$

$$N_{x\theta} = -(L-x)p\cos\theta$$

13.19

Now the cylinder length does not change:

$$\int_{-L/2}^{L/2}(N_x - \nu N_\theta)dx = 0$$

Substituting Eqs. (b) of Example 13.6 into this, taking $f_1 = 0$ and integrating the resulting expression, we have

$$f_2(\theta) = \nu\gamma a^2(1-\cos\theta) - \frac{\gamma L^2}{24}\cos\theta$$

Referring to Example 13.6, the solution is thus,

$$N_\theta = \gamma a^2(1-\cos\theta)$$

$$N_{x\theta} = \gamma ax\sin\theta \quad ◄$$

$$N_x = \frac{\gamma x^2}{2}\cos\theta + \nu\gamma a^2(1-\cos\theta) - \frac{\gamma L^2}{24}\cos\theta$$

13.20

Referring to Fig. 13.12b:

$$p_x = p\sin\phi \qquad p_z = p\cos\phi$$

$$r_o = x\cos\phi$$

<u>Stress resultants due to weight</u>:

$$F = 2\pi r\cdot p\sin\phi\cdot r d\phi$$

Since

$$r d\phi = dx, \qquad r = x\cot\phi$$

Then,

$$F = 2\pi\int_0^x x\cot\phi\, p\sin\phi\, dx$$

$$= 2\pi p\cos\phi\left(\frac{x^2}{2}\right) + c$$

For a cone supported at ts edge c=0, since F=0 at x=0. Therefore, Eq. (13.46b) gives

$$N_\phi = -\frac{px}{2\sin\phi} \qquad (1)$$

Equation (13.46a):

$$N_\theta = \frac{p_z r_o}{\sin\phi} = -px\frac{\cos^2\phi}{\sin\phi} \qquad (2)$$

<u>Stress resultants due to pressure</u>:

Equation (13.46a) yields,

$$N_\theta = -p_r\frac{x\cos\phi}{\sin\phi} = -p_r x\cot\phi \qquad (3)$$

We now have

$$F = (2\pi r\, p_r\sin\phi\, r d\phi)\cos\phi$$

Following a procedure similar to that the preceding, we obtain

$$F = 2\pi p_r\cos^2\phi\left(\frac{x^2}{2}\right)$$

Equation (13.46a) leads to

$$N_\phi = -\frac{p_r x\cos\phi}{2\sin\phi}$$

$$= -\frac{1}{2}p_r x\cot\phi \qquad (4)$$

Solution is determined by the superposition of the preceding results: adding Eqs. (4) and (1), and (3) and (2). In so doing, we have

$$N_x = -\frac{x}{2\sin\phi}(p + p_r\cos\phi) \quad ◄$$

and

$$N_\theta = -x\cot\phi\left(p\cos\phi + \frac{1}{2}p_r\right) \quad ◄$$